Firefighter Fatalities
in the
United States
in 1999

Prepared for

National Fire Data Center
United States Fire Administration
Federal Emergency Management Agency
Contract No. EME-1998-CO-0202-T0006

Prepared by

IOCAD Emergency Services Group

July 2000

Table of Contents

ACKNOWLEDGMENTS

This study of firefighter fatalities would not have been possible without the cooperation and assistance of many members of the fire service across the United States. Members of individual fire departments, chief fire officers, the National Interagency Fire Center, United States Forest Service personnel, the United States military, the Department of Justice, the National Fire Protection Association, and many others contributed important information for this report.

IOCAD Emergency Services Group of Emmitsburg, Maryland (a division of IOCAD Technical Services, Inc.) conducted this analysis for the United States Fire Administration (USFA) under contract EME-1998-CO-0202-T0006.

The ultimate objective of this effort is to reduce the number of firefighter deaths through an increased awareness and understanding of their causes and how they can be prevented. Firefighting, rescue, and other types of emergency operations are essential activities in an inherently dangerous profession, and unfortunate tragedies do occur. This is the risk all firefighters accept every time they respond to an emergency incident. However, the risk can be reduced greatly through efforts to increase firefighter health and safety.

The USFA would like to extend its thanks to photographer Kathleen Cole and to the *Chattanooga Times and Free Press* for permission to use the photograph on the cover of this report. The photograph shows the funeral procession for Fire and Rescue Captain Lewis Edward Williams who died on May 14, 1999, at the scene of a trench rescue in Fort Oglethorpe, Georgia.

This report is dedicated to the families of those firefighters who made the ultimate sacrifice in 1999. May the lessons learned from their passing not go unheeded.

BACKGROUND

For more than 20 years, the United States Fire Administration (USFA) has tracked the number of firefighter fatalities and conducted an annual analysis. Through the collection of information on the causes of firefighter deaths, the USFA is able to focus on specific problems and direct efforts towards finding solutions to reduce the number of firefighter fatalities in the future. This information also is used to measure the effectiveness of current programs directed toward firefighter health and safety.

In addition to the analysis, the USFA provides a list of firefighter fatalities to the National Fallen Firefighters Foundation. If Memorial criteria are met, the fallen firefighter's next of kin, as well as members of the individual fire department, are invited to the annual Fallen Firefighters Memorial Service. The Service is held at the National Emergency Training Center in Emmitsburg, Maryland annually during Fire Prevention Week. Additional information regarding the Memorial Service can be found on the Internet at http://www.firehero.org/ or by calling the National Fallen Firefighters Foundation at (301) 447-1365. An updated list of firefighter fatalities from 1981 through the present, including a searchable database, can be found at http://www.usfa.fema.gov/ffmem/ffmem_search.cfm

INTRODUCTION

This report continues a series of annual studies by the USFA of on-duty firefighter fatalities in the United States.

The specific objective of this study was to identify all of the on-duty firefighter fatalities that occurred in the United States in 1999, and to analyze the circumstances surrounding each occurrence. The study is intended to help identify approaches that could reduce the number of firefighter deaths in future years.

In addition to the 1999 overall findings, this study includes special analyses on vehicle collisions and personal protective clothing and equipment use.

Who is a Firefighter?

For the purpose of this study, the term firefighter covers all members of organized fire departments, including career and volunteer firefighters; full-time public safety officers acting as firefighters; State and Federal government fire service personnel, including wildland firefighters; and privately employed firefighters, including employees of contract fire departments and trained members of industrial fire brigades, whether full- or part-time. It also includes contract personnel working as firefighters or assigned to work in direct support of fire service organizations.

Under this definition, the study includes not only local and municipal firefighters, but also seasonal and full-time employees of the United States Forest Service, the Bureau of Land Management, the Bureau of Indian Affairs, the Bureau of Fish and Wildlife, the National Park Service, and State wildland agencies. The definition also includes prison inmates serving on firefighting crews; firefighters employed by other governmental agencies such as the United States Department of Energy; military personnel performing assigned fire suppression activities; and civilian firefighters working at military installations.

What Constitutes an On-Duty Fatality?

On-duty fatalities include any injury or illness sustained while on-duty that proves fatal. The term on-duty refers to being involved in operations at the scene of an emergency, whether it is a fire or non fire incident; responding to or returning from an incident; performing other officially assigned duties such as training, maintenance, public education, inspection, investigations, court testimony, and fund-raising; and being on call, under orders, or on stand by duty, except at the individual's home or place of business. An individual who experiences a heart attack or other fatal injury at home as he or she prepares to respond to an emergency is considered on-duty when the response begins.

A fatality may be caused directly by an accidental or intentional injury in either emergency or non emergency circumstances, or it may be attributed to an occupationally related fatal illness. A common example of a fatal illness incurred on-duty is a heart attack. Fatalities attributed to occupational illnesses also would include a communicable disease contracted while on-duty that proved fatal, where the disease could be attributed to a documented occupational exposure.

Injuries and illnesses are included where death is delayed considerably after the original incident. When the incident and the death occur in different years, the analysis counts the fatality as having occurred in the year that the incident occurred.

There is no established mechanism for identifying fatalities that result from illnesses that develop over long periods of time, such as cancer, which may be related to occupational exposure to hazardous materials or products of combustion. It has proved to be very difficult over several years to provide a full evaluation of an occupational illness as a causal factor in firefighter deaths, because of the limitations in the ability to track the exposure of firefighters to toxic hazards, the often delayed long-term effects of such exposures, and the exposures firefighters may receive while off-duty.

Sources of Initial Notification

As an integral part of its ongoing program to collect and analyze fire data, USFA solicits information on firefighter fatalities directly from the fire service and from a wide range of other sources. These sources include the Public Safety Officers' Benefit Program (PSOB) administered by the Department of Justice, the National Institute for Occupational Safety and Health (NIOSH), the Occupational Safety and Health Administration (OSHA), the United States military, the National Interagency Fire Center, and other Federal agencies.

The USFA receives notification of some deaths directly from fire departments, as well as from such fire service organizations as the International Association of Fire Chiefs (IAFC), the International Association of Fire Fighters (IAFF), the National Fire Protection Association (NFPA), the National Volunteer Fire Council (NVFC), State fire marshals, State training organizations, other State and local organizations, fire service Internet sites, news services, and fire service publications. The USFA also keeps track of fatal fire incidents as part of its Major Fire Investigations Program and performs an ongoing analysis of data from the National Fire Incident Reporting System (NFIRS).

Procedure for Including a Fatality in the Study

In most cases, after notification of a fatal incident, initial telephone contact is made with local authorities by the USFA's contractor to verify the incident, its location and jurisdiction, and the fire department or agency involved. Further information about the deceased firefighter and the incident may be obtained from the chief of the fire department or his or her designee over the phone or by other data collection forms.

Information that is requested routinely includes NFIRS-1 (incident) and NFIRS-3 (fire service casualty) reports, the fire department's own incident reports and internal investigation reports, copies of death certificates or autopsy results, special investigative reports, police reports, photographs and diagrams, and newspaper or media accounts of the incident. Information on the incident also may be gathered from the NFPA, USFA, or NIOSH reports on an incident.

After obtaining this information, a determination is made as to whether the death qualifies as an on-duty firefighter fatality according to the previously described criteria. The same criteria were used for this study as in previous annual studies. Additional information may be requested, either by follow up with the fire department directly, from State vital records offices, or other agencies. The determination as to whether a fatality qualifies as an on-duty death for inclusion in this statistical analysis is made by the USFA. The final determination as to whether a fatality qualifies as a line-of-duty death for inclusion in the Fallen Firefighters Memorial Service is made by the National Fallen Firefighters Foundation.

1999 FINDINGS

One hundred and twelve (112) firefighters died while on-duty in 1999. This represents an increase of 21 deaths from 1998, and reverses the overall downward trend from the levels of deaths that had been experienced during the decade. The total of 112 fatalities is the highest number recorded since 119 firefighters gave their lives in 1989. The lowest years on record were 1992, with 75 fatalities, and 1993, with 77 fatalities.

This year's total reverses the long-term downward trend of reduced fatalities that began in 1979, after a peak of 171 in 1978. Despite a horrible year in 1999, the overall trend in firefighter fatalities is down 24 percent over the last 10 years. However, the rate of reduction in the last 5 years has slowed to 6 percent, attributable partly to the uncharacteristically low number of deaths that occurred in 1992 and 1993 (Figure 1). The total of 112 for 1999 is 16 percent higher than the 10-year and 5-year trends.

Figure 1 - On-Duty Firefighter Fatalities (1977-1999)

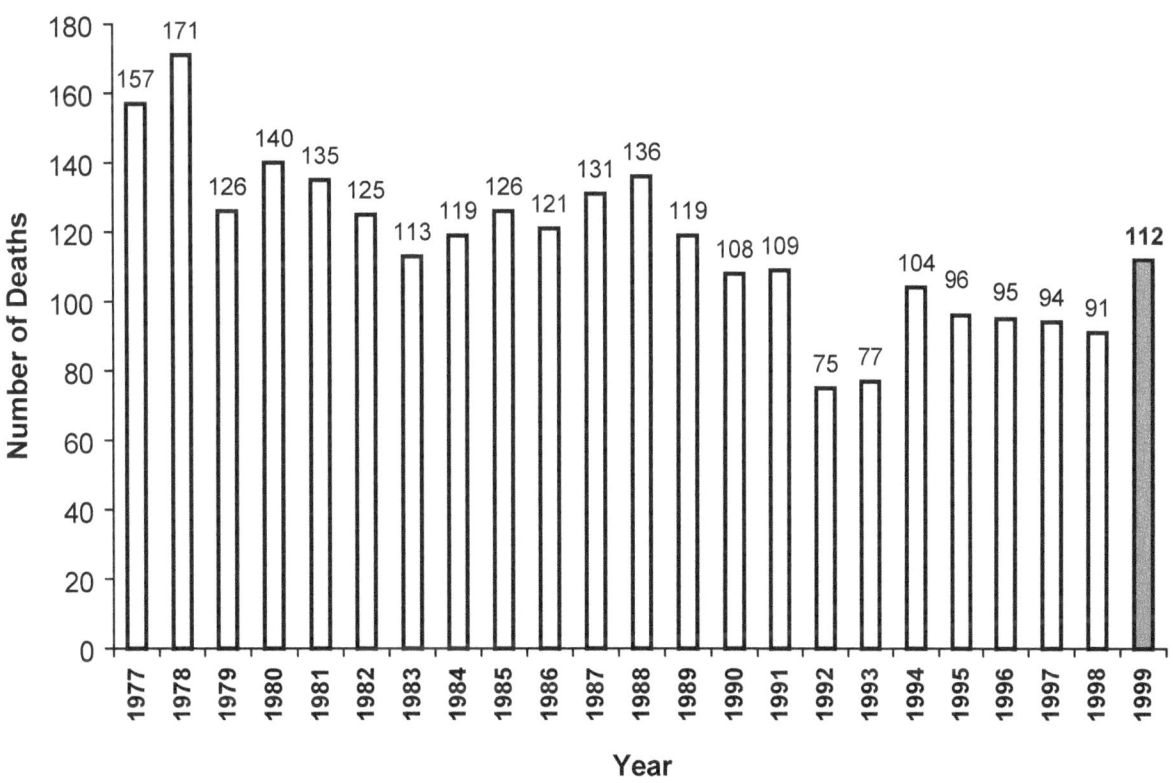

The 1999 firefighter fatalities included 74 volunteer firefighters and 38 career firefighters (Figure 2). Among the volunteer firefighter fatalities, 71 were from local or municipal volunteer fire departments and 3 were seasonal or contract members of wildland fire agencies. Of the career firefighters who died, 37 were members of local or municipal fire departments and 1 was a wildland career firefighter. One hundred and ten (110) of the fatalities were men and 2 were women.

Figure 2
Career vs. Volunteer Deaths

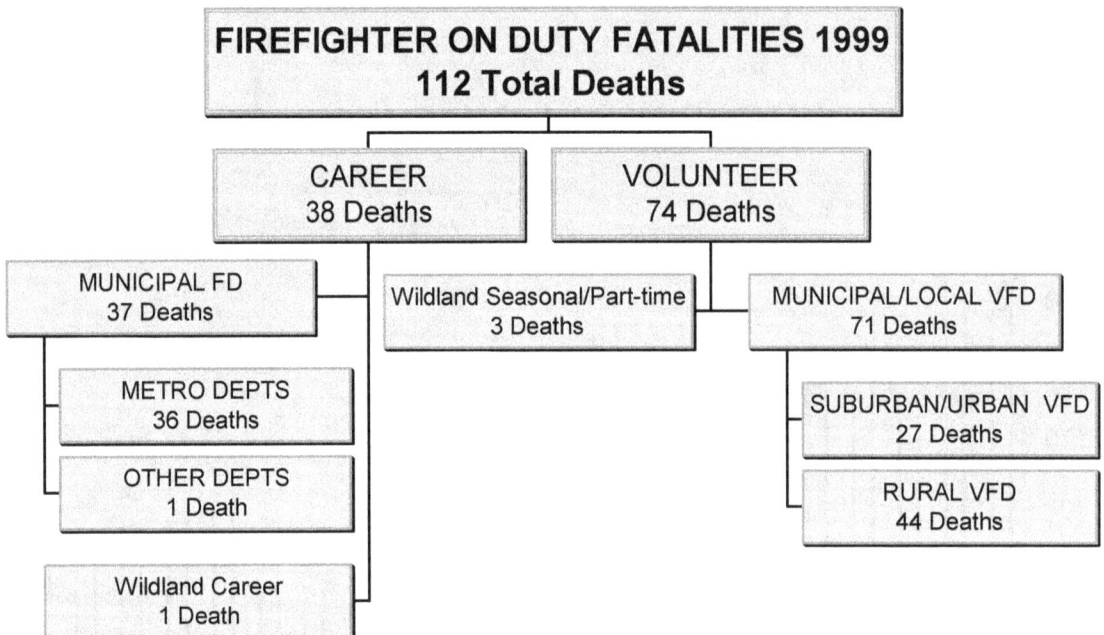

The 112 deaths resulted from 100 incidents. Six (6) multi fatality incidents resulted in the deaths of 18 firefighters; this is down from the total of 10 multi fatality incidents that occurred in 1998, which resulted in 22 firefighter deaths. In 1999, 6 Massachusetts firefighters were killed when they became disoriented in a six-story cold storage warehouse fire; 3 Texas firefighters were killed in the collapse of a fire-involved church attic; 3 Iowa firefighters were killed in the flashover of an apartment building as they performed search and rescue activities; 2 Kentucky firefighters died when they were overcome by fire progress in a wildland fire; 2 District of Columbia firefighters were killed when they were overcome by fire progress in a residence; and 2 Indiana firefighters died after they were injured in the rollover of a fire department tanker (tender).

The number of deaths associated with brush, grass, or wildland firefighting rose to 28. This is a sharp rise from the 13 experienced in 1998, the 9 deaths experienced in 1997, and the 5 deaths experienced in 1996. 1999 also reflects a significant increase over the 18 firefighters who died in wildland activities in 1995. For the first time since 1993 and for only the third time since 1970, there were no firefighter fatalities resulting from aircraft crashes. At least 1 firefighting aircraft crash occurred-- a helicopter in California--but the pilot's life was spared.

Two (2) Kentucky firefighters were killed when they were overcome by fire progress while building a fire break in hardwood leaf litter on the floor of a wooded area; 2 Indiana firefighters were injured and later died as the result of a tanker rollover while responding to a brush fire; 11 firefighters died of heart attacks in separate brush or wildland incidents; 3 firefighters were killed when they were electrocuted in separate brush or wildland incidents; 3 firefighters were killed after being struck or run over by fire apparatus in separate incidents; 2 firefighters were killed in tanker rollovers while en route to wildland fires in separate incidents; 1 firefighter drowned at a woods fire; 1 firefighter died of heat stroke as he served as an EMT at a wildland fire; 1 firefighter fell 150 feet to his death at a wildland fire; 1 firefighter was struck and killed by a falling boulder at a wildland fire; and, 1 firefighter experienced a fatal cerebrovascular accident (CVA) while driving a rescue vehicle to a wildland fire.

In 1999, 4 of the 28 wildland-related firefighter deaths involved vehicle rollovers where the firefighter was the driver or an occupant of the vehicle. There were 3 such incidents, all of which involved unrestrained firefighters being ejected from the vehicle and killed. The fact that many firefighters fail to use seatbelts will be examined later in this report.

Type of Duty

In 1999, 97 on-duty firefighter deaths were associated with emergency incidents, accounting for 87 percent of the 112 fatalities (Figure 3). This includes all firefighters who died while responding to an emergency, while at the emergency scene, or while returning from the emergency incident. Non emergency activities accounted for 15 fatalities (13 percent). Non emergency duties include training, administrative activities, or performing other functions that are not related to an emergency incident. A historic perspective concerning the percentage of firefighter deaths that occur during emergency duty is presented in Table 1.

Figure 3 - Firefighter Deaths by Type of Duty (1999)

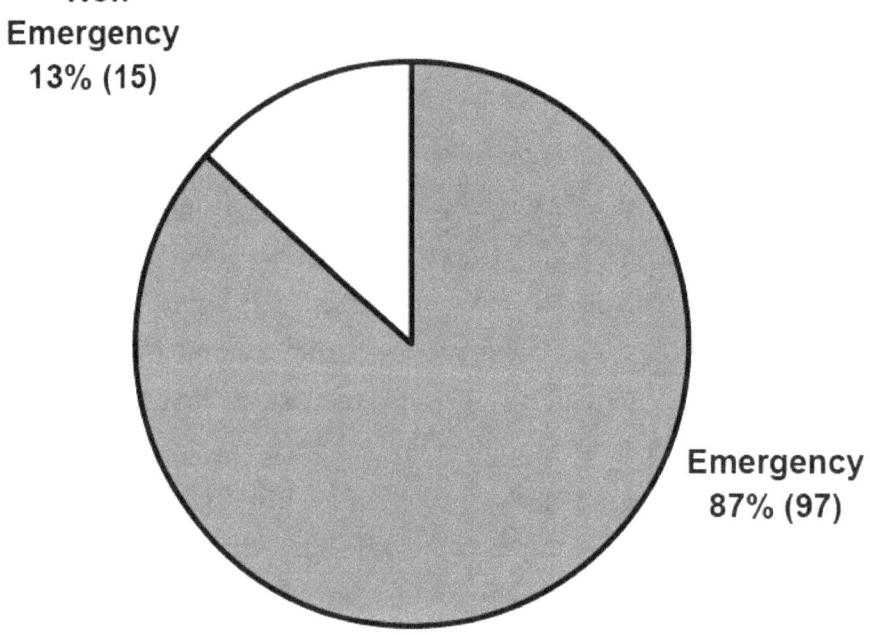

Non Emergency 13% (15)

Emergency 87% (97)

Table 1. Emergency Duty Firefighter Deaths	
Year	**Percent**
1999	87%
1998	77%
1997	81%
1996	72%
1995	85%

The number of deaths by type of duty being performed in 1999 is shown in Table 2 and presented graphically in Figure 4. As in previous years, the largest number of deaths occurred during fireground operations. There were 60 fireground deaths, which accounted for 54 percent of the fatalities, up from 46 percent in 1998 and 44 percent in 1997. Of the 60 fireground deaths, more than 2 in 5 (25) resulted from heart attacks that occurred on the fire scene. Other fireground deaths included 14 from asphyxiation, 8 from burns, 8 from internal trauma, and 3 electrocutions. One (1) firefighter died of heat stroke as he worked as an EMT on a wildland fire incident and 1 firefighter drowned at the scene of a woods fire.

Table 2. Type of Duty – 1999	Number	Percent
Fireground Operations	60	54%
Responding/Returning from Alarm	26	23%
Other On-duty	11	10%
Non Fire Emergencies	7	6%
After an Incident	5	4%
Training	3	3%
TOTAL	112	100%

Figure 4 - Fatalities by Type of Duty (1999)

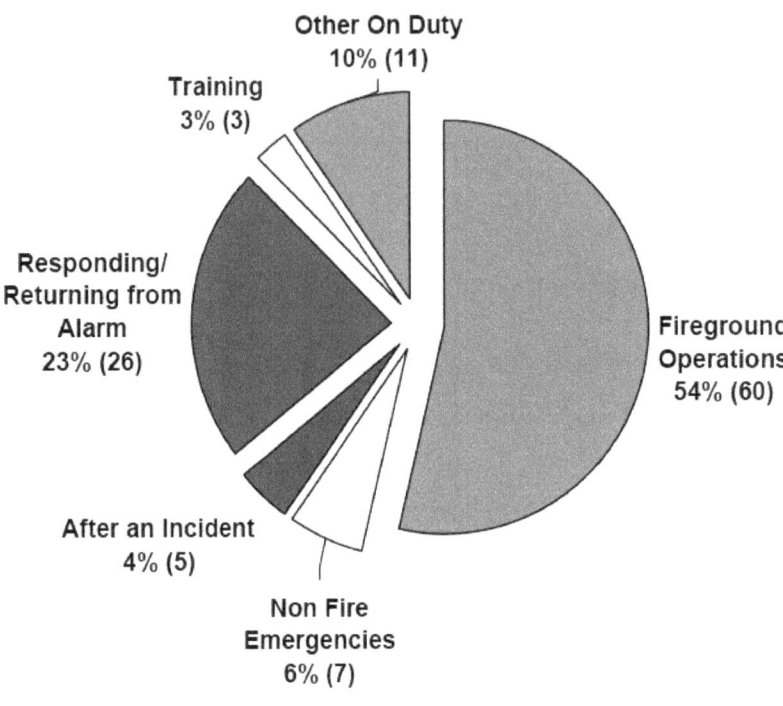

Twenty-six (26) firefighters died while responding to or returning from emergency incidents. This has been the second leading type of duty in which firefighter deaths have occurred each year since 1993. The number of firefighters killed performing this type of duty is up sharply from 1998 when 14 firefighters were killed, but similar to the number of deaths that occurred in 1995 when 29 firefighters died. In 1999, 23 of these deaths involved volunteer firefighters.

Of the 26 firefighters who died while responding to or returning from alarms, 13 died of heart attacks, 2 died of strokes or CVA's, and 11 died from traumatic injuries. Two (2) firefighters died in collisions involving their personal vehicles, 1 firefighter was killed when he was ejected through the windshield of a pumper after a collision, 1 firefighter was killed when the rescue truck in which he was a passenger hit a tree, 1 firefighter fell from a ladder truck as it responded to an automatic fire alarm, and a total of 6 firefighters were killed in collisions, most of them rollovers.

There were 11 deaths that occurred during other on-duty activities. These deaths include 5 firefighters who died from heart attacks while on-duty and 2 who died from pulmonary embolisms. One (1) firefighter was killed when a part of a front-end loader that he was using to clear snow from fire hydrants failed and struck the firefighter, 1 firefighter was killed in an explosion of fireworks residue as he supervised a controlled burn, 1 firefighter suffered a stroke or CVA as he was engaged in a physical fitness workout, and 1 firefighter died of heart inflammation while on-duty.

Seven (7) deaths were related to activities at the scene of non fire emergency incidents; this is half of the number reported in 1998. 2 of the deaths involved fire police officers, 1 died of a heart attack, and 1 died after being struck by a tractor trailer truck as he directed traffic. Two (2) firefighters died of heart attacks while engaged in vehicle extrication incidents, 1 firefighter died of a heart attack as he aided an injured high school football player, 1 firefighter died of a heart attack while engaged in a trench rescue incident, and 1 firefighter was killed when he was struck by a car at the scene of an earlier motor vehicle collision.

Five (5) firefighters died in 1999 after the conclusion of an incident. Three (3) deaths were heart attacks, 1 was a stomach aneurysm, and the fifth was a fire investigator who was killed when he was crushed by the collapse of a chimney in the attic of a building where he was conducting an investigation.

In 1999, 3 firefighters died during training exercises. This number is significantly lower than the 12 that were experienced in 1998 and more in line with the experience of recent years. In 1997, 5 firefighters died while training, 6 died in 1996, and the total for 1995 was 3. The firefighter deaths in 1999 included 1 firefighter who drowned during a dive training exercise, 1 firefighter who collapsed and died of a heart attack during a physical fitness test, and 1 firefighter who was killed as a result of head and facial trauma after attempting an emergency egress procedure during a training exercise.

Career, Volunteer, and Wildland Deaths by Type of Duty

Figure 5 depicts career, volunteer, and wildland firefighter deaths by type of duty. Wildland career, wildland seasonal, and wildland contractor deaths were grouped together. This chart demonstrates the disproportionate number of fatalities experienced by volunteer firefighters responding to and returning from alarms as compared to career and wildland firefighters. This is a continuing trend. The large number of career firefighter deaths while on-duty but not involved in an incident or training activity, may be attributed to the fact that career firefighters are on-duty for longer periods of time than volunteer firefighters. The on-duty periods for volunteer firefighters generally are related to an emergency incident or other official functions such as training. Some volunteer fire departments staff stations overnight (similar to a career department) but their number is small when compared to the total number of volunteer fire departments.

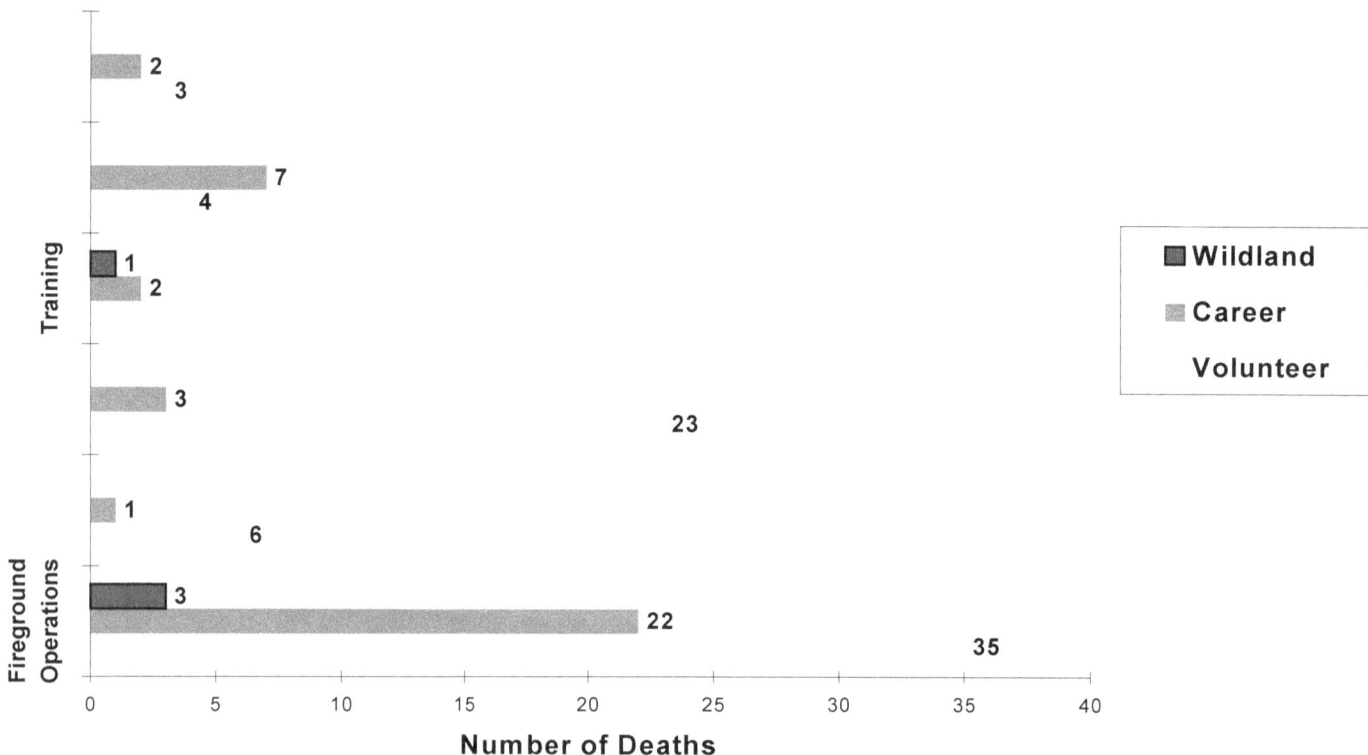

Figure 5 - Career, Volunteer, and Wildland Deaths by Type of Duty (1999)

Type of Emergency Duty

Ninety-three (93) firefighters died while engaged directly in emergency service delivery, including deaths that were the result of injuries sustained on the incident scene or en route to the incident scene (4 died immediately after an emergency incident). Figure 6 shows the percentage of firefighters killed in firefighting, emergency medical services, hazardous materials, false alarms, and technical rescue-related incidents. Seventy-six (76) firefighters were killed in relation to fires, 12 in relation to EMS calls, 1 associated with a hazardous materials emergency, and 1 was killed while engaged in a technical rescue. The firefighters killed in relation to false alarms consisted of 1 firefighter who experienced a heart attack at a call that involved the prank activation of a fire alarm pull station, 1 firefighter who died of a heart attack at the scene of a fire alarm activation caused by carpet installation activity, and 1 firefighter who fell from a ladder truck en route to what turned out to be a false alarm.

Figure 6 - Type of Emergency Duty (1999)

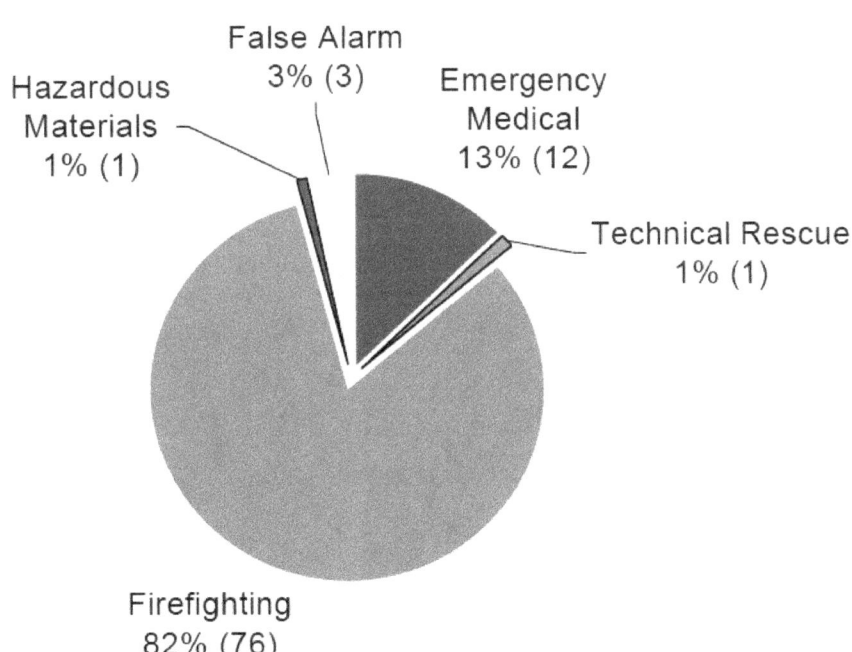

Note - 93 of 112 – On-Scene or Responding

Cause of Fatal Injury

As used in this study, the term "cause of injury" refers to the action, lack of action, or circumstances that resulted directly in the fatal injury, while the term "nature of injury" refers to the medical cause of the fatal injury or illness, often referred to as the physiological cause of death. A fatal injury usually is the result of a chain of events, the first of which is recorded as the cause. For example, if a firefighter is struck by a collapsing wall, becomes trapped in the debris, runs out of air before being rescued, and dies of asphyxiation, the cause of the fatal injury is recorded as "struck by collapsing wall" and the nature of the fatal injury is "asphyxiation." Similarly, if a wildland firefighter is overrun by a fire and dies of burns, the cause of the death would be listed as "caught/trapped," and the nature of death would be "burns." This follows the convention used in the National Fire Incident Reporting System (NFIRS) casualty reports.

Figure 7 shows the distribution of deaths by cause of fatal injury or illness; Table 3 presents the exact number.

Figure 7 - Fatalities by Cause of Fatal Injury (1999)

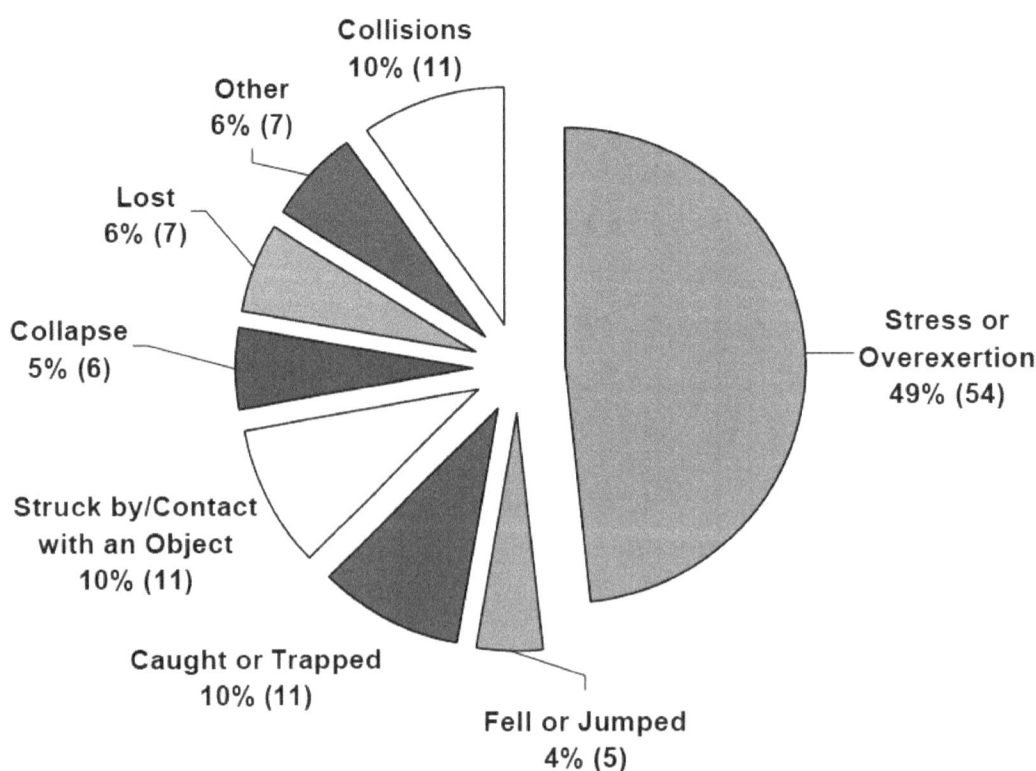

Table 3. Cause of Fatal Injury - 1999	Number	Percent
Stress or Overexertion	54	49%
Collisions	11	10%
Caught or Trapped	11	10%
Struck By/Contact with an Object	11	10%
Lost	7	6%
Other	7	6%
Collapse	6	5%
Fell or Jumped	5	4%
	112	100%

As in most previous years, the largest cause category is stress or overexertion, which was listed as the primary factor in 49 percent of the deaths, up from 46 percent in 1998, and 43 percent in 1997, but returning to the 50 percent level that has been experienced in the last several years. Firefighting is extremely strenuous physical work and is likely one of the most physically demanding activities that the human body performs. Most firefighter deaths attributed to stress result from heart attacks. Of the 54 stress-related fatalities in 1999, 52 firefighters died of heart attacks and 2 died of CVA's. Seven (7) of the 54 deaths for which the cause is listed as stress/overexertion occurred during non emergency activities.

Eleven (11) firefighters were killed in collisions. Two (2) Indiana firefighters died as the result of injuries that were received in a tanker rollover, 1 firefighter was ejected through the windshield of a pumper after a collision, 2 firefighters were killed in separate crashes involving their personal vehicles during a response, 1 firefighter was killed as he rode in the officer's seat of a heavy rescue that crashed into a tree while responding to a hazardous materials incident, and 5 firefighters died in separate crashes involving tankers or pumpers.

Tied for the second leading cause of firefighter fatalities was being caught or trapped, which also accounted for 11 deaths (10 percent), down from 18 percent in 1998. These deaths included 3 multi fatality incidents: 1 in Washington, DC where 2 firefighters were caught by fire progress, one in Keokuk, Iowa, where 3 firefighters were trapped in a flashover that occurred as they searched an apartment for fire victims; and, 1 in Cranston, Kentucky, where 2 firefighters were overcome by fire progress as they built a fire break in a forest fire. An Illinois firefighter was killed in an electrical explosion, a Missouri firefighter drowned during a training exercise, an Arkansas firefighter was burned as he prepared a hose line for use on a fire involving a recreational vehicle when the fuel tank failed and surrounded him with fire, and a New York firefighter was killed in an explosion as he supervised the disposal of fireworks residue.

Also tied for the second leading cause of firefighter fatalities for 1999 was being struck by or coming into contact with an object. There were 11 deaths in this category, including 3 firefighters who were killed in separate electrocutions, all on brush or wildland fires. Two (2) firefighters acting as fire police officers were killed when they were struck by passing vehicles, 1 firefighter was struck by a passing vehicle as he filled a fire department tanker, 1 firefighter was struck by a car that skidded out of control as he worked on the scene of an earlier collision, 1 firefighter died after being hit by a falling rock on a wildland fire, 1 firefighter was killed when he was struck by a part of a loader that he was repairing, and 2 firefighters were struck and killed by fire department vehicles on the scene of wildland fires.

Seven (7) firefighters died when they became lost inside burning structures. This total includes the 6 firefighters killed in a cold storage warehouse in Massachusetts and a Missouri firefighter killed when he became disoriented in a warehouse fire.

Six (6) firefighters were killed in collapses. Three (3) Texas firefighters died while fighting a church fire when the roof collapsed and trapped them, a New York firefighter was killed when a chimney collapsed and crushed him as he investigated a fire, a California firefighter was killed in a structural collapse, and an Indiana firefighter was killed when he was trapped by a roof collapse in a residential fire.

Five (5) firefighters were killed when they fell or jumped. This total includes a Pennsylvania firefighter who fell from a ladder truck as it responded to a call, a California firefighter who attempted an emergency egress method and fell to the ground, a Pennsylvania firefighter who fell through a floor at a structure fire and received a fatal back injury, a wildland firefighter who fell 150 feet to his death, and a California firefighter who fell or jumped off a pumper as it began to move and was crushed by its rear wheels.

Seven (7) firefighters were killed in incidents that are not otherwise classified. These incidents include a Washington, DC, firefighter who was attacked and injured by a dog and later died, a Tennessee firefighter who drowned at a woods fire, a New York firefighter who ran out of air in a basement fire, a firefighter who died of a heart disease not related to stress, 1 firefighter who experienced a CVA (stroke), and 2 firefighters who suffered pulmonary embolisms while on-duty.

Nature of Fatal Injury

Table 4 and Figure 8 show the distribution of the 112 deaths by the medical nature of the fatal injury or illness. The leading nature of death in 1999 was heart attacks, which accounted for 52 deaths, up from 38 firefighter fatalities in 1998, and 36 in 1997.

Table 4. Nature of Fatal Injury	Number	Percent
Heart Attacks	52	46%
Internal Trauma	25	22%
Asphyxiation (includes drowning)	16	14%
Burns	8	7%
CVA/Stroke	3	3%
Electrocution	3	3%
Other	3	3%
Pulmonary Embolism	2	2%
TOTAL	112	100%

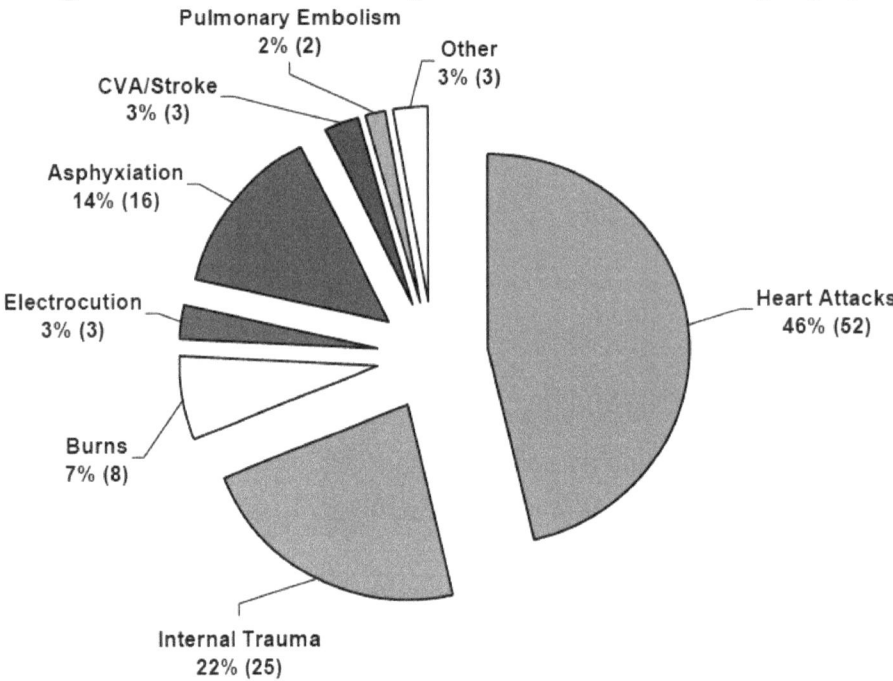

Figure 8 - Fatalities by Nature of Fatal Injury (1999)

Figure 9 provides a detailed breakdown of heart attacks by type of duty. Twenty-five (25) of the heart attacks occurred at the fire scene and 13 occurred while en route to or returning from an emergency incident. This represents a major increase from 1998, when 17 and 6, respectively, died. One (1) occurred at a training incident, down from 7 in 1998, and 5 occurred during non-fire incidents, the same as in 1998. Five (5) heart attacks occurred during other on-duty times and 3 occurred just after the firefighter responded to an incident.

Figure 9 - Heart Attacks by Type of Duty (1999)

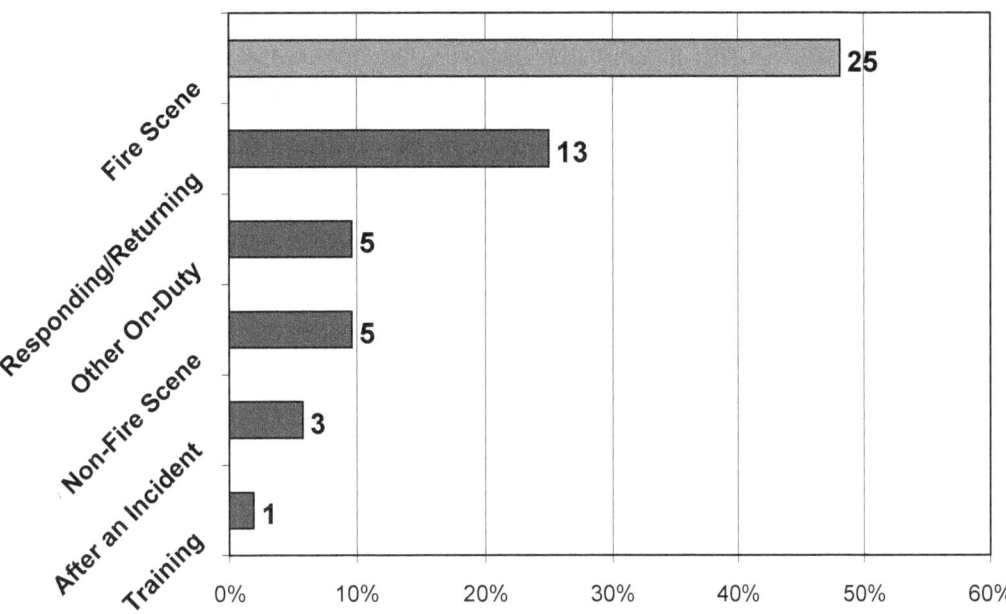

Internal trauma was the second leading nature of death, responsible for 25 deaths (down from 27 in 1998, and 32 in 1997). This total includes 1 firefighter who was crushed by a falling chimney, 8 firefighters who were killed in fire apparatus collisions, 2 firefighters who were killed in collisions involving their personal vehicles, 2 fire police officers who were struck by vehicles, 4 firefighters who were killed in falls, 2 firefighters who were struck and killed by fire department vehicles, 2 firefighters who were struck by non fire department vehicles, 1 firefighter who was struck and killed by falling rocks, 1 firefighter who was killed in an explosion as he supervised the disposal of fireworks residue, 1 firefighter who was struck by a piece of equipment that he was repairing, and a firefighter who fell through a floor during a structure fire.

Asphyxiation was the third leading medical reason for firefighter deaths, responsible for 16 deaths (1 more than the totals for the two previous years). Four (4) incidents claimed 10 firefighters: 4 of the Worcester firefighters, 2 in Kentucky, 2 of the Keokuk firefighters, and 2 of the Texas firefighters killed while fighting a church fire. Two (2) firefighters drowned, a California firefighter was trapped by a structural collapse, a Missouri firefighter became disoriented in a warehouse fire, a New York firefighter ran out of air in a basement fire, and an Indiana firefighter was trapped by a roof collapse in a residential fire. Twelve

(12) of these deaths occurred during structural firefighting, 2 occurred in a wildland fire, and 2 were drownings.

Eight (8) of the 112 firefighter fatalities that occurred in 1999 were attributed to burns. This total included 2 of the Worcester firefighters, 2 District of Columbia firefighters, 2 of the firefighters killed in Keokuk, Iowa, 1 of the firefighters killed in a Texas church fire, an Illinois firefighter killed as the result of an electrical explosion, and an Arkansas firefighter killed while fighting a fire in a recreational vehicle.

Three (3) firefighters were killed by strokes (CVA's), one of which occurred in the fire station as a firefighter completed a physical fitness workout. One (1) firefighter suffered a stroke as he drove a pumper to a fire or explosion involving an electrical transformer; the pumper left the road and crashed into a concrete culvert but the injuries he received in the crash were not severe. The third firefighter suffered a stroke as he responded to a brush fire.

Three (3) firefighters died from electrocution. One (1) firefighter tripped or fell as he advanced a booster line and fell face first onto a live line, 1 firefighter came into contact with a fence that had been energized by a power line, and 1 firefighter was killed as he stomped out some burning embers that were found in an alley; however, unbeknownst to anyone, an electrical line had fallen and was hidden by brush.

Two (2) on-duty firefighters died from pulmonary embolisms.

The nature of 3 other firefighter deaths were heat stroke while serving as an EMT on a wildland fire, heart inflammation while on-duty, and a stomach aneurysm after an incident.

Firefighters' Ages

Figure 10 shows the distribution of firefighter deaths by age and cause of death. Younger firefighters were more likely to have died as a result of traumatic injuries from an apparatus accident or after becoming caught or trapped during firefighting operations. Stress was shown to play an increasing role in firefighter deaths as age increased.

Figure 10 - Fatalities by Age & Cause (1999)

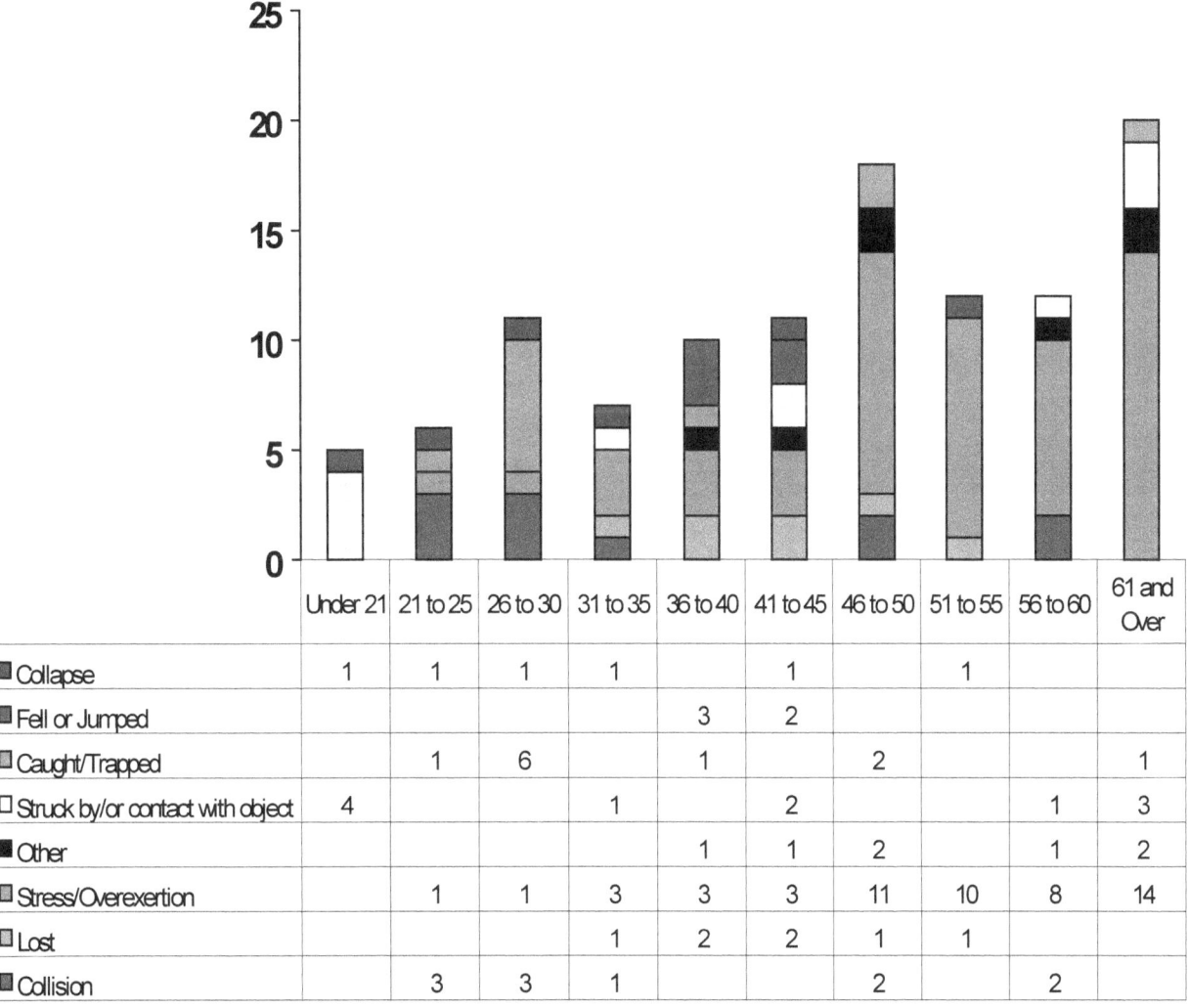

	Under 21	21 to 25	26 to 30	31 to 35	36 to 40	41 to 45	46 to 50	51 to 55	56 to 60	61 and Over
■ Collapse	1	1	1	1		1		1		
■ Fell or Jumped					3	2				
▨ Caught/Trapped		1	6		1		2			1
□ Struck by/or contact with object	4			1		2			1	3
■ Other					1	1	2		1	2
▨ Stress/Overexertion		1	1	3	3	3	11	10	8	14
▨ Lost				1	2	2	1	1		
▨ Collision		3	3	1			2		2	

This also is reflected in Figure 11, which shows the distribution of deaths by age and medical nature of injury. Trauma, electrocution, and asphyxiation were responsible for most of the deaths of younger firefighters, while heart attacks were much more prevalent among older firefighters. Heart attacks accounted for 65 percent of firefighter deaths where the firefighter was age 41 or higher.

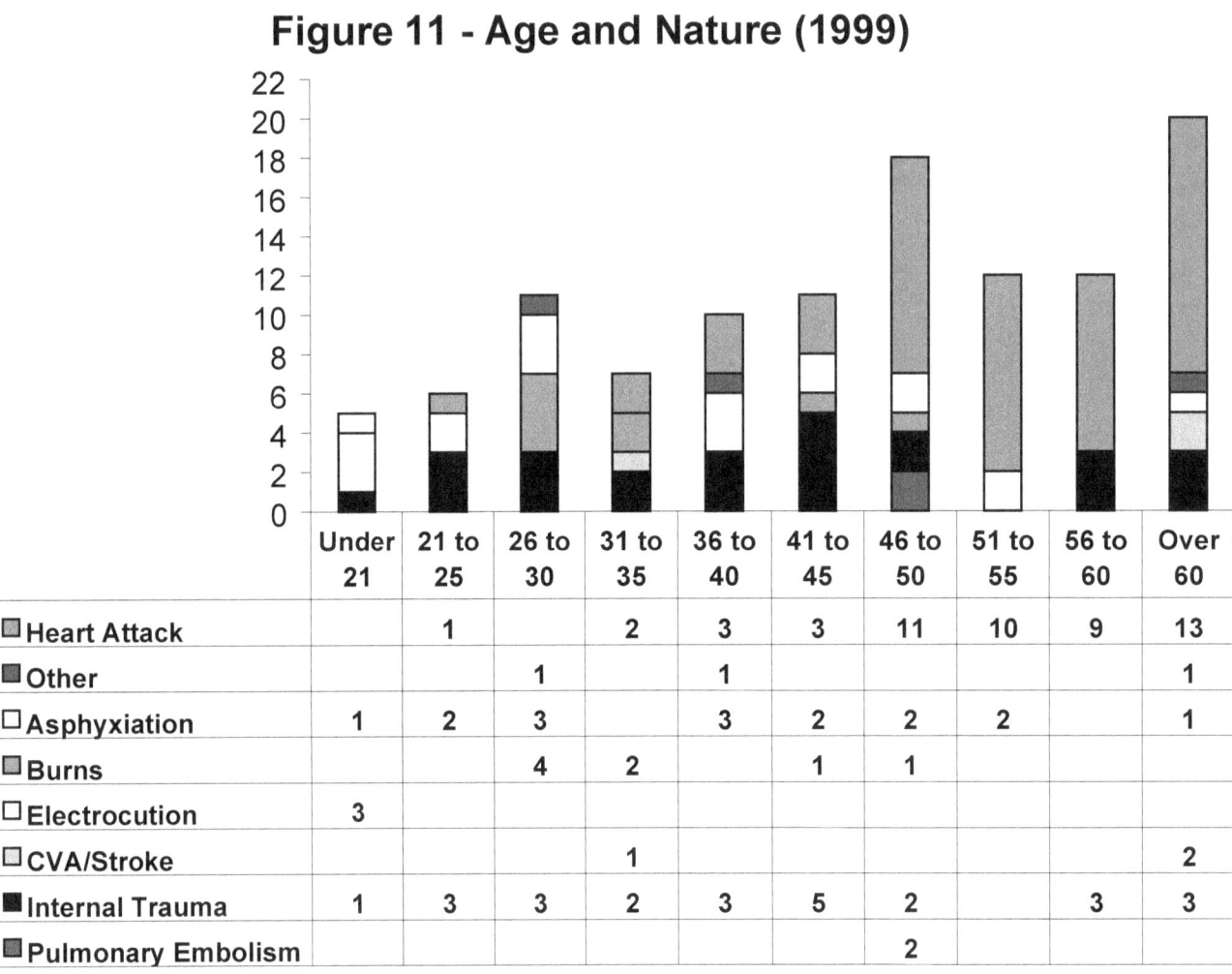

Figure 11 - Age and Nature (1999)

	Under 21	21 to 25	26 to 30	31 to 35	36 to 40	41 to 45	46 to 50	51 to 55	56 to 60	Over 60
Heart Attack		1		2	3	3	11	10	9	13
Other			1		1					1
Asphyxiation	1	2	3		3	2	2	2		1
Burns			4	2		1	1			
Electrocution	3									
CVA/Stroke				1						2
Internal Trauma	1	3	3	2	3	5	2		3	3
Pulmonary Embolism							2			

Fixed Property Type

There were 60 fireground deaths in 1999. Table 5 and Figure 12 show the distribution by fixed property use.

Table 5. Property Use for Fireground Deaths	Number	Percent
Residential	23	38%
Outdoor Property	16	27%
Storage	7	12%
Commercial	6	10%
Institutional	4	7%
Street/Road	3	5%
Public Assembly	1	2%
TOTAL	60	101%*

* - Total greater than 100% due to rounding

Figure 12 – Fatalities by Fixed Property Use

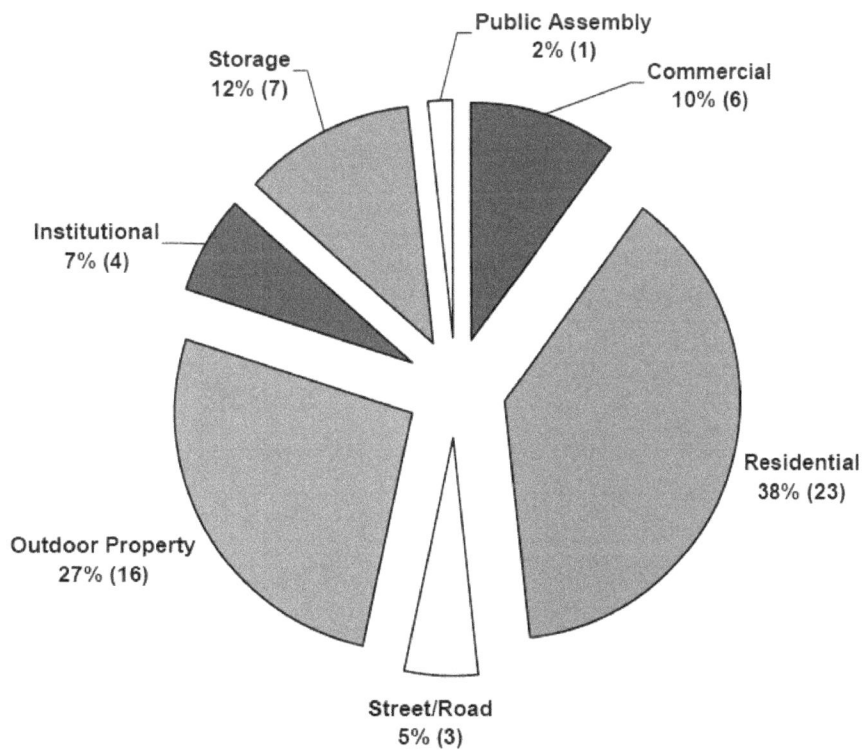

Fireground Deaths

Structural fires accounted for 41 fireground deaths. As in most years, residential occupancies accounted for the highest number of these fireground fatalities, with 23 deaths (more than half of all structural fire deaths). Residential occupancies usually account for 70 to 80 percent of all structure fires and a similar percentage of the civilian fire deaths each year[1]. The frequency of firefighter deaths in relation to the number of fires is much higher for non-residential structures.

Fires that occurred on outdoor properties and "street/road" accounted for a total of 19 deaths.

Type of Activity

Table 6 and Figure 13 show the type of fireground activity that the 60 firefighters were engaged in at the time they sustained their fatal injuries or illnesses.

Table 6. Type of Activity for Fireground Deaths	Number	Percent
Advancing Hose Lines/Fire Attack	16	27%
Search and Rescue	10	17%
Water Supply	8	13%
Other	5	8%
Incident Command	4	7%
Cutting Fire Breaks (Wildland)	4	7%
Support Duties	4	7%
Investigating	3	5%
Salvage/Overhaul	3	5%
Scene Safety	3	5%
TOTAL	60	101%*

*- Total greater than 100% due to rounding.

---

[1] Complete 1999 NFIRS fire incidence data were not available at the time of this report, but residential fires typically account for between 70 and 80 percent of all civilian fatalities each year.

Figure 13 - Fatalities by Type of Activity (1999)

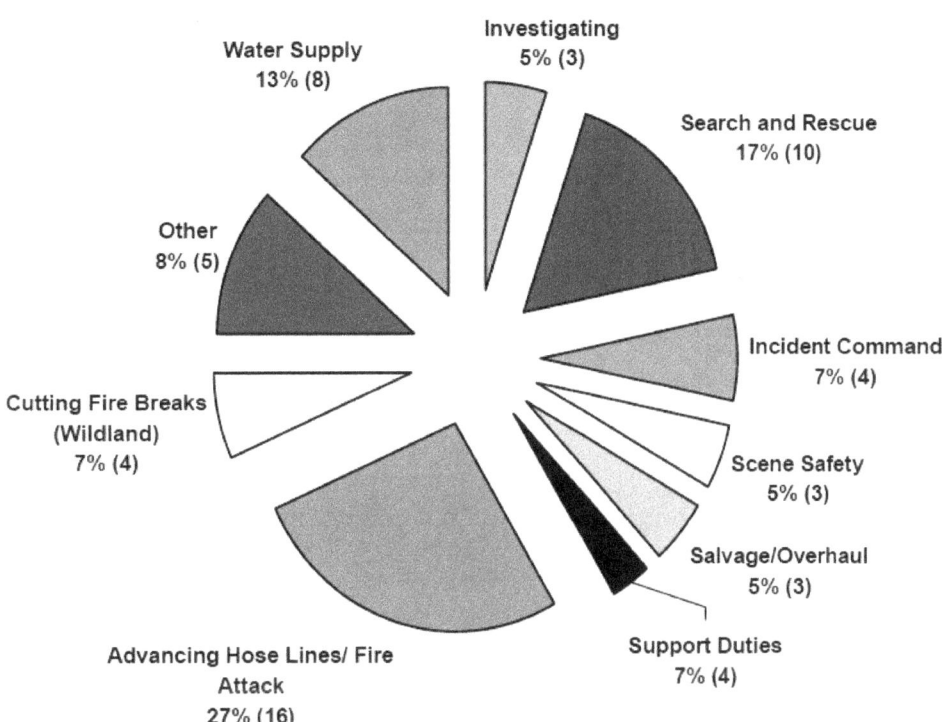

Sixteen (16) firefighters died while engaged in fire attack and advancing hose lines. This is a small but significant decrease compared to the 18 firefighters killed in 1998 during such activity and the 21 firefighters killed performing this activity in 1997. Eight (8) of the 16 firefighters killed were engaged in structural firefighting, 7 died at brush or wildland fires, and 1 died at a vehicle fire.

Ten (10) firefighters were killed during search and rescue operations (up three from the level in 1998 and five from the levels experienced in 1997 and 1996). Six (6) were lost in a warehouse in Massachusetts as they searched for fire victims and lost firefighters, 3 died in an apartment fire in Iowa as they searched for fire victims, and 1 died in a basement fire in New York when he ran out of air while searching.

Eight (8) firefighters were killed while engaged in water supply operations (twice the level of 1998 and 1997). Seven (7) of the eight were heart attacks: 3 at structure fires, 2 at vehicle fires; and, 2 at brush fires. One (1) firefighter was killed when he was struck by a passing vehicle as he was filling a tanker in support of a structural firefighting operation.

Four (4) firefighters died while performing incident command activities, 3 of heart attacks and 1 of asphyxiation as he performed command duties in the interior of a warehouse fire.

Four (4) firefighters died while cutting fire breaks, twice the number of 1998. Two (2) wildland firefighters died in California in different incidents; 1 was struck by a boulder and 1 fell 150 feet to his death. Two (2) Kentucky firefighters were killed when they were overcome by fire progress in a woods fire.

Four (4) firefighters died while performing support duties at fires, one up from in 1998. 3 deaths occurred at structural fires. One (1) firefighter died of heat stroke as he performed Emergency Medical Technician (EMT) duty at a wildland fire in California.

Three (3) firefighters died while investigating fires or fire scenes. One (1) firefighter in Illinois was killed in an electrical explosion, a District of Columbia firefighter was attacked and later died after being injured by a dog as he investigated a report of fire, and a Massachusetts firefighter died from a heart attack experienced at a report of a fire.

Three (3) firefighters died while performing salvage or overhaul duties; all three were at structural fires and all three were heart attacks.

Three (3) firefighters died while performing scene safety duties; 2 of heart attacks and 1 after being struck by a vehicle.

Five (5) firefighters died in circumstances not classified above, including a firefighter who drowned at a woods fire, a firefighter who suffered a heart attack at the scene of a nursing home fire, 2 firefighters who were killed at wildland fire scenes when they were struck by fire apparatus, and a Texas firefighter who suffered a heart attack after he and other firefighters discovered a fire while off duty (he was considered on-duty from the moment the fire was discovered).

Time of Injury

The distribution of all 1999 firefighter deaths according to the time of day when the fatal injury occurred is illustrated in Figure 14 (nine incidents were not reported by time of day).

Figure 14 - Fatalities by Time of Fatal Injury (1999)

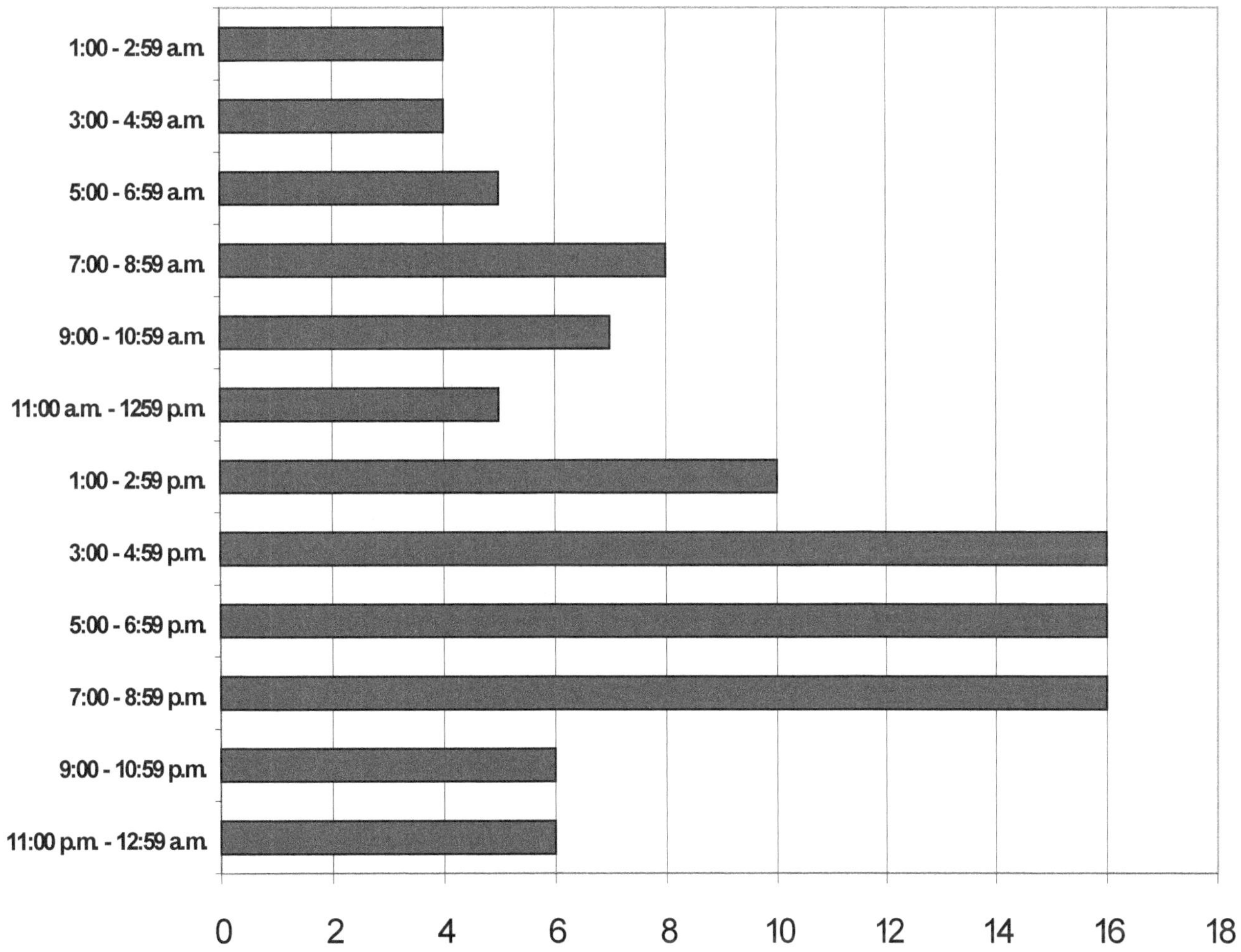

Month of the Year

Figure 15 illustrates firefighter fatalities by month of the year. Firefighter fatalities peaked in December with 19 deaths. December 1999 was the worst month for firefighter deaths since before 1994. Other than the fact that wildland fires and their associated injuries and deaths occur in the wildland season, no trends were identified.

Figure 15 - Deaths by Month of the Year (1999)

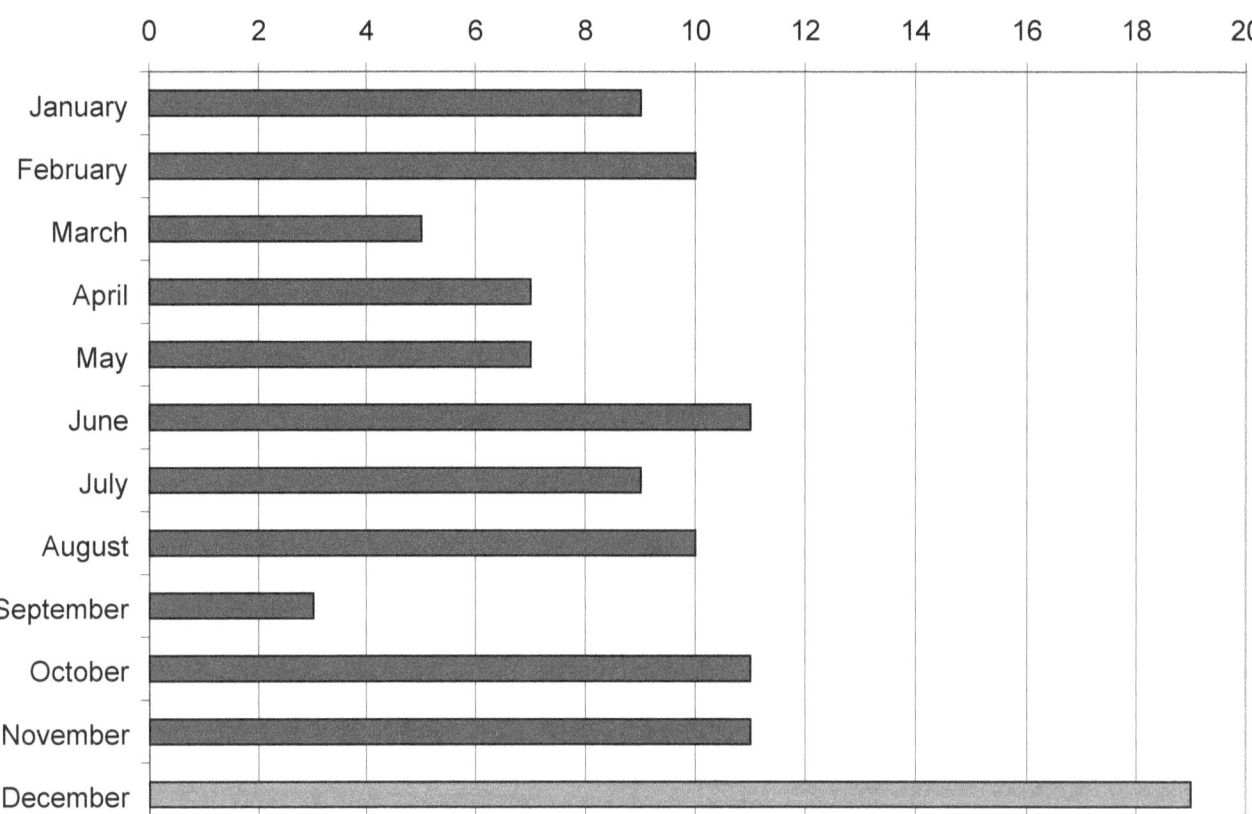

State and Region

The distribution of firefighter deaths by State is shown in Table 7.[2] Thirty-two States and the District of Columbia each had at least 1 firefighter fatality. Pennsylvania had the highest number of deaths with 13 followed by Massachusetts with 12. Figure 16 shows the firefighter fatalities divided by region of the country and their status as career, volunteer, or wildland firefighters.

Figure 16.
Firefighter Deaths By Region 1999

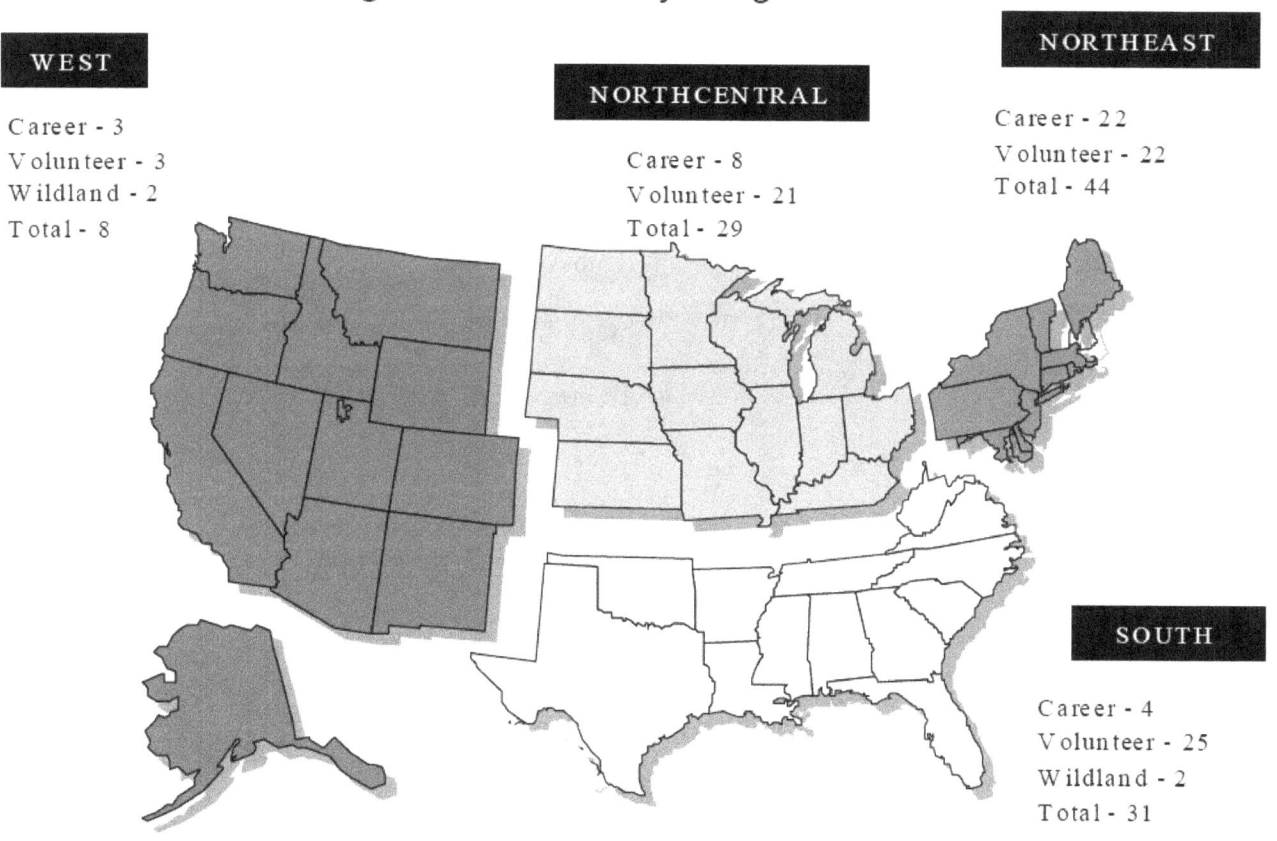

WEST

Career - 3
Volunteer - 3
Wildland - 2
Total - 8

NORTHCENTRAL

Career - 8
Volunteer - 21
Total - 29

NORTHEAST

Career - 22
Volunteer - 22
Total - 44

SOUTH

Career - 4
Volunteer - 25
Wildland - 2
Total - 31

---

[2] This list attributes the deaths according to the State in which the fire department or unit is based, as opposed to the State in which the death occurred. They are listed by those States for statistical purposes, and for the National Fallen Firefighters Memorial at the National Fire Academy.

Table 7. 1999 States with On-duty Firefighter Fatalities

State	Number of Deaths	State	Number of Deaths
Alabama	1	Mississippi	2
Arkansas	3	Missouri	6
California	6	Nebraska	1
Connecticut	1	New Jersey	4
District of Columbia	3	New Mexico	1
Florida	1	New York	8
Georgia	2	North Carolina	1
Illinois	3	Ohio	7
Indiana	5	Oklahoma	2
Iowa	3	Oregon	1
Kentucky	2	Pennsylvania	13
Louisiana	3	South Carolina	3
Maryland	2	Tennessee	1
Massachusetts	12	Texas	10
Michigan	1	Vermont	1
Minnesota	1	Virginia	1
		West Virginia	1

Total - 112

Analysis of Urban/Rural/Suburban Patterns in Firefighter Fatalities

The United States Bureau of the Census defines "urban" as a place having a population of at least 2,500 or lying within a designated urban area. Rural is defined as any community that is not urban. Suburban is not a census term but may be taken to refer to any place, urban or rural, that lies within a metropolitan area defined by the Census Bureau, but not within one of the central cities of that metropolitan area.

Fire department areas of responsibility do not always conform to the boundaries used for the census. For example, fire departments organized by counties or special fire protection districts may have both urban and rural coverage areas. In such cases, it may not be possible to characterize the entire coverage area of the fire department as rural or urban, and firefighter deaths were listed as urban or rural based on the particular community or location in which the fatality occurred.

The following patterns were found for 1999 firefighter fatalities. These statistics are based on answers from the fire departments and, when no data from the department were available, the data are based upon population and area served reported by the fire departments.

Table 8.	Urban/Suburban	Rural	Federal or State Parks/Wildland	Total
Firefighter Deaths	62	45	5	112

SPECIAL TOPICS

Vehicle Collisions Involving Firefighters

In 1999, 14 firefighters died as the result of collisions involving fire apparatus, collisions involving firefighters' personal vehicles, or by falling from fire apparatus. A short synopsis of each incident follows.

- Firefighter Kenneth Alan Strain (age 28) of the Hembly Bridge Volunteer Fire Department in North Carolina was killed on May 2, 1999, when he lost control of a pumper while returning from an EMS call. Firefighter Strain pulled the pumper to the right to allow traffic to pass him when he lost control, was unable to regain control, and hit a tree. Firefighter Strain was wearing a seatbelt.

- Firefighter Cilton Jay Dauzat (age 63) of the White Tail Ridge Volunteer Fire Department in Texas was killed on August 5, 1999, when he fell or slipped from the back step of a pumper and was run over as the pumper backed down a hill.

- Firefighter Timmy Ray Dawson (age 34) of the Center Rock Volunteer Fire Department in South Carolina was killed on August 31, 1999, when the right wheels of a pumper he was driving left the roadway. Firefighter Dawson overcompensated for this condition and the pumper left the right side of the road and rolled several times. Speed was a factor in this incident. Firefighter Dawson was not wearing a seat belt. Two (2) other firefighters who were passengers in the vehicle received minor injuries; they were wearing seatbelts.

- Firefighter Kenneth C. Cashman (age 29) of the Auglaize Township Volunteer Fire Department in Ohio died on September 13, 1999, when his personal vehicle was struck by a dump truck during a response. Firefighter Cashman was wearing a seatbelt.

- Captain William Malcolm Bethune (age 58) of the Texas City Fire Department in Texas was killed on October 5, 1999, when he was ejected through the windshield of his pumper. The pumper had been involved in a collision while responding to an EMS call and struck a bridge support. Captain Bethune was not wearing a seatbelt.

- Firefighter Elvis Benson Maxwell (age 49) of the Grant Parish Fire District #5 in Louisiana was killed on October 7, 1999, when he lost control of the tanker he was driving. Firefighter Maxwell was not wearing a seatbelt.

- Firefighter Karen Jane Savage (age 44) of the Hawkins Bar Volunteer Fire Department in California was killed on October 16, 1999, when she fell or jumped from a pumper that was beginning to move and was crushed by the pumper's rear wheels.

- Firefighter Brian K. Burnett (age 23) and Captain Robert Charles Ulrich (age 57) of the Scipio Township Volunteer Fire Department in Indiana were injured and subsequently died as the result of the failure of the tanker to negotiate a curve during a response on October 28, 1999. The pumper left the roadway and rolled several times. Firefighter Burnett was not wearing a seatbelt and was ejected and crushed by the tanker as it rolled.

- Firefighter Michael J. Sims, Sr. (age 38) of the Highland Hose Company in Tarentum, Pennsylvania, was killed on November 2, 1999, when he fell from a ladder truck during a response. The apparatus was equipped with seatbelts and a safety gate. It is unknown if Firefighter Sims used either the seatbelt or the safety gate.

- Firefighter David Zan Lancaster (age 24) of the Elliott Volunteer Fire Department in Mississippi was killed on November 7, 1999, when he was involved in a collision while driving his personal vehicle on a response. It is unknown whether Firefighter Lancaster was wearing a seatbelt.

- Firefighter Bert Andrew Bruecher (age 46) of the Village of Pleak Volunteer Fire Department in Texas was killed on November 14, 1999, when the tanker he was driving failed to negotiate a curve. The tanker left the roadway and rolled. Firefighter Bruecher was partially ejected and pinned under the truck.

- Firefighter Paul Franklin Ezernack, Jr. (age 28) of the North Sabine Fire Protection District in Louisiana was killed on December 15, 1999, when an embankment gave way under the right wheels of the tanker he was driving. Firefighter Ezernack attempted to regain control but the tanker left the roadway and rolled. Firefighter Ezernack was thrown 170 yards.

- Firefighter Bradley Curtis McNeer (age 22) of the Chesterfield County Fire Department in Virginia was killed on December 18, 1999, when the rescue truck in which he was a passenger collided with a tree. The driver had diverted his attention from the roadway, veered off the right side of the road, overcompensated as he brought the truck back onto the road, and overcompensated again to avoid collisions with other vehicles on the roadway. He then struck a tree on the right side of the road. Firefighter McNeer was wearing a seat belt.

These deaths illustrate the need for firefighters riding on or driving fire apparatus to wear their seatbelts, and the need for fire apparatus drivers to be given training on the proper procedure to be used if the right wheels of a vehicle leave the roadway while the vehicle is in motion.

SeatBelts

Seatbelts are the most basic piece of safety equipment provided with a motor vehicle. In 1999, a number of drivers and apparatus occupants were killed when they were ejected from fire apparatus that was involved in a collision or rollover. Fire departments must take steps to assure that seatbelts are used by all firefighters. Some suggested steps for assuring their use include.

- All vehicles operated by the fire department must be equipped with seatbelts. Older vehicles that may not be equipped with belts may be retrofitted by the manufacturer.

- Each fire department should have written policies that require seat belt use by all emergency vehicle drivers and passengers. This requirement must be strictly enforced in all vehicle use situations, including response to emergencies, returning from emergencies, and all other driving.

- Firefighters must remain in their seatbelts throughout the entire emergency response. It is unacceptable to leave the protection of a seatbelt to don protective clothing or self-contained breathing apparatus (SCBA) while the vehicle is moving. In addition, unrestrained SCBA's can become projectiles in a collision and must be mounted in positive latching brackets if carried in the cab. Other items carried in the cab, like flashlights, clipboards, and map books also must be secured.

- Signs or placards should be placed in each fire department vehicle to remind firefighters of seatbelt use requirements.

- A seatbelt must be present and used by each vehicle occupant. If there are more seats than seatbelts, vehicle capacity should be limited to the number of seated, belted positions.

- Riding on the exterior of the vehicle, at any stage of the emergency, should be expressly forbidden in fire department Standard Operating Procedures (SOP's). There is no safe place to ride on the exterior of an emergency vehicle. The only exception allowed by NFPA 1500, *Standard on Fire Department Occupational Safety and Health Program*, is for hose loading. There are a number of controls that are required by the standard.

Regaining Control of the Vehicle

As can be seen from the incident descriptions on the previous pages, a very common scenario is for the right wheels of the apparatus to leave the road for some reason, for the driver to fight to regain control of the vehicle, and for the vehicle to collide with a fixed object such as a tree or to roll over when the apparatus leaves the left side of the road.

Fire apparatus are wide vehicles and fire apparatus drivers/operators may have a tendency to hug the right side of the road. The following guidelines can help to reduce the incidence of this type of collision or to reduce the severity of such collisions.

- Fire apparatus drivers/operators must limit their speed to one that is reasonable for the road conditions, the size of the vehicle, the configuration of the vehicle, the topography of the response route (curves, hills), and visibility (fog, dark, precipitation). Many fire departments have adopted speed limits for responding apparatus that are a maximum of 10 miles per hour above the posted speed limit **if conditions permit**. In some areas or States, fire apparatus may not be permitted to travel above the speed limit in any situation. Many of the collisions described in this section had speed as a contributing factor.

- Younger or inexperienced drivers demand special attention to reinforce the importance of speed control and decisions about the need to respond safely.

- If possible, avoid small roads with no improved shoulder. This is not possible or practical in all situations, especially emergency response. If a driver/operator has the opportunity to do so, small roads with unimproved shoulders should be avoided.

- The apparatus driver/operator should use as much of the road as is needed to operate safely. Especially during an emergency response where traffic is yielding, a position closer to the center of the road will prevent slipping off to the right. In some situations, such as in curves and on hills, the best place for the responding apparatus is in their defined lane at a slower speed.

- If a vehicle leaves the right side of the road and conditions permit, the driver should take his or her foot off of the throttle and slow to a speed that allows the vehicle to come back onto the roadway safely. "Fighting" to get the vehicle back on the road at speed may lead to a collision or rollover.

- A publication available from the USFA, *"Emergency Vehicle Driver Training,"* provides excellent information on driver safety for emergency vehicles. The publication FA-110 is free and can be obtained through the internet as a PDF document or as hard copy at http://www.usfa.fema.gov/usfapubs or by writing to:

   United States Fire Administration
   Publications Center
   16825 South Seton Avenue
   Emmitsburg, MD 21727

- Additional emergency vehicle operations information may be found from the following sources. A cost or charge for the material may be involved.

    Emergency Vehicle Response Safety
    VFIS
    183 Leader Heights Road
    P.O. Box 2726
    York, PA 17405
    (800) 233-1957
    http://www.vfis.com/vfis/cets/seminar.htm

    Safe Passage
    National Volunteer Fire Council
    1050 17th Street, NW
    Suite 1212
    Washington, DC 20036
    (888) 275-6832
    http://www.nvfc.org

Personal Protective Clothing and Equipment

In 1999, 16 firefighters died in situations where the provision and use of personal protective clothing possibly could have prevented their deaths. It is impossible to express their chances of survival in absolute terms; however a discussion of these situations is appropriate so that we may learn from these experiences. The following firefighter deaths did or likely involved the absence, lack of use, inappropriate use, or failure of personal protective clothing and equipment:

- On February 16, 1999, Firefighter Robert C. Stanmire (age 52) of the Forest Grove Volunteer Fire Company of Vineland, New Jersey, died of a heart attack suffered while he was responding to a brush fire. Firefighter Stanmire was wearing full structural protective clothing.

- On March 16, 1999, Firefighter Charles James "Chuck" Vodak (age 45) died of a heart attack after fighting a prairie fire for more than 4 four hours in structural firefighting clothing.

- On April 6, 1999, Captain Kenneth Allen Nickell (age 28) and Firefighter/EMT Kevin Rex Smith (age 30) of the Route 377 Volunteer Fire Department in Kentucky were killed when they were overcome by fire progress in a woods fire. Both firefighters were equipped with brush firefighting protective clothing, but were not equipped with personal fire shelters.

- On May 30, 1999, Firefighter Lewis Jefferson Matthews (age 29) of the District of Columbia Fire Department was killed when he and other firefighters were overcome by fire progress in a residential fire. Firefighter Matthews was equipped with a manually activated Personal Alarm Safety System (PASS) device that was found in the "off" mode. Another firefighter trapped by the same fire had an automatically activated PASS device that sounded and aided rescuers in his discovery. Despite being removed from the fire area sooner than Firefighter Matthews, the other firefighter also died.

- On August 8, 1999, Firefighter Ronald Wade Meshell (age 30) of the Huttig Fire Department in Arkansas was killed while fighting a fire in a recreational vehicle (RV). Firefighter Meshell was downhill from the RV when the vehicle's fuel tank failed, surrounding Firefighter Meshell with a pool of burning fuel. Firefighter Meshell had loaded his personal protective clothing on the truck but was not wearing it at the time. He was burned over 96 percent of his body surface.

- On December 3, 1999, 6 firefighters of the Worcester Fire Department in Massachusetts were killed when they became disoriented and lost in a smoky warehouse fire. The presence or status of PASS devices for these firefighters is unknown at this time. NIOSH is investigating the incident and may provide more information on this subject in its report.

- On December 20, 1999, Battalion Chief John H. Tvedten (age 47) of the Kansas City, Missouri, Fire Department died when he became lost in a warehouse fire. The presence or status of his PASS device is unknown at this time.

- On December 22, 1999, Assistant Chief David M. McNally (age 48), Firefighter Jason L. Bitting (age 29), and Firefighter Nathan R. Tuck (age 39) of the Keokuk Fire Department in Iowa were killed in a flashover in an apartment fire. All 3 firefighters wore PASS devices that were integrated with their SCBA's. The PASS devices failed to sound but their failure is not considered a factor in these deaths.

These deaths point to the need for each firefighter to be provided with, and to use, personal protective clothing and equipment that is appropriate for the hazard. The following guidelines may be used to assist fire department managers and firefighters:

- Firefighters must be provided with personal protective equipment (PPE) that is appropriate for the hazard. Brush gear and fire shelters provide appropriate protection in wildland fire situations. This type of equipment is inexpensive when compared to structural firefighting protective clothing and sometimes may be procured at discounted prices from State forestry departments and the Federal wildland fire programs. Structural firefighting clothing is not appropriate for wildland fires. The stress placed on a firefighter by the ambient atmosphere, as well as the weight and confines of structural protective clothing, are excessive for wildland circumstances.

- Firefighters must wear protective clothing and activate protective equipment items which are available to them. Firefighter Meshell had his protective clothing with him but was not wearing it; and a number of firefighters were wearing PASS devices that were not activated.

- SCBA must be worn in all situations that are or may become Immediately Dangerous to Life and Health (IDLH). This includes vehicle fires.

- Protective equipment failures, such as the one in Keokuk, Iowa, must be investigated by trained professionals. Clothing and equipment that have been involved in a failure incident should be impounded, stored in paper bags or cardboard boxes to prevent corrosion, and examined by someone trained to do so. NIOSH is a resource that has conducted a number of studies of SCBA's that have experienced failures.

CONCLUSIONS

The analysis of firefighter deaths in 1999 indicates that the overall long-term trend toward fewer firefighter fatalities has suffered a step backward. The 112 deaths that occurred in 1999 represent a level that is 16 percent higher than the trend over the last decade. Firefighters continue to provide more services to their communities; the addition of emergency medical services provision, hazardous materials response, and technical rescue service has brought risks and accompanying firefighter deaths.

There were 28 firefighter deaths associated with brush, grass, or wildland fires in 1999, a marked increase from the experience of the past 4 years. This total is despite the fact that there were no deaths caused by wildland firefighting aircraft crashes. Three (3) of these deaths were from electrocutions, all involving firefighters under the age of 21. Three (3) firefighters died on fire scenes after being run over by fire apparatus, and a total of 4 firefighters died after being occupants of fire apparatus that rolled while responding to brush fires. Fires involving vacant land sometimes may be viewed as routine incidents. The unfortunate brush, grass, and wildland death toll of 1999 should point to the fact that there is no such thing as a routine incident. One of the special topics in this report addresses appropriate protective clothing for these types of incidents. Firefighters can be overprotected and/or underprotected for these types of incidents if the proper equipment is not provided.

A disproportionate number of volunteer firefighters are being killed while responding to and returning from alarms. Almost one third of all volunteer firefighter deaths occurred while performing this type of duty. Of the 23 volunteer firefighters who died performing this type of duty in 1999, 11 died of heart attacks, 10 died in vehicle collisions, including 4 who died in rollovers and 2 who died in collisions involving their personal vehicles, 2 died of CVA's (strokes), and 1 died in a fall. Six (6) of the 10 volunteer firefighters who died in vehicle collisions were under the age of thirty. One of the special topics of this report addresses safety during emergency response.

In December 1999, the eyes of the world were on Worcester, Massachusetts. Six (6) firefighters died after 2 firefighters became disoriented in a windowless structure while searching for fire victims and called for help. Other firefighters entered the building to search for the lost firefighters and subsequently gave their lives. Their deaths reminded the world of the hazards of firefighting. The month of December 1999 was the worst in terms of firefighter fatalities in a number of years with a total of 19 firefighters dead in 12 incidents, including the tragic deaths of 3 firefighters in Keokuk, Iowa three days before Christmas.

Stress-induced heart attacks remained the top cause of firefighter deaths. Continued focus on firefighter health and wellness likely may reduce the impact of this killer in the future.

Continued attention to basic health maintenance, physical fitness, periodic medical exams, the use of seatbelts, emergency vehicle operator training, PASS devices, SCBA's, and other protective clothing and equipment items will be the cornerstone for future efforts to reduce this loss of life.

Although there is hope that the future brings with it less tragedy, even 1 firefighter death brings tragedy and hurt to families and communities. One hundred (100) communities suffered this supreme loss in 1999.

APPENDIX A
SUMMARY OF 1999 INCIDENTS

If additional information is available regarding a firefighter fatality, the reader is directed to these sources. Where possible, hyperlinks that direct the reader to additional information are provided. If links have expired or if the reader does not have Internet access, contact information for these sources is provided at the end of the appendix when available.

1/5/99
Carl Arnold Olsen, Firefighter
Age 51, Volunteer
Kiln Volunteer Fire Department, Mississippi

Firefighter Olsen was killed when he was struck by a car as he worked to refill a water tanker (tender). The incident occurred at approximately 7 p.m. The driver of the car was blinded by the lights of the tanker. Firefighter Olsen was not wearing any reflective material at the time he was struck. Firefighter Olsen's wife is also a volunteer firefighter and was on the scene at the time of the fatal incident.

1/8/99
Daniel J. O'Connell, Firefighter
Age 58, Volunteer
Putnam Lake Fire Department, New York

Firefighter O'Connell responded with his department to a report of a structure fire. The fire involved an oil burner in a furnace and was extinguished with the use of a dry chemical extinguisher. Firefighter O'Connell, who had some experience with this type of equipment, was ordered to investigate further and make sure that the fire was out. He was exposed to soot and dry chemical residue. He left the house and collapsed on the front lawn in cardiac arrest. EMS treatment was begun immediately and Firefighter O'Connell was transported to the hospital. It is the opinion of Firefighter O'Connell's doctor that the heart attack was caused by exposure to smoke and chemicals. Firefighter O'Connell suffered from injuries to the brain caused by a lack of oxygen, and died on January 25, 1999. No autopsy was performed.

1/9/99
Jason A. Gouckenour, Firefighter
Age 22, Volunteer
Worthington-Jefferson Volunteer Fire Department, Indiana

Firefighter Gouckenour entered a structural fire in a house alone with a hose line. He was equipped with full turnout gear and an SCBA, but was not equipped with a PASS device. It is believed that he tripped over a coffee table and became entangled in a couch. He removed his SCBA to call for help and was overcome by extremely heavy heat and smoke conditions. Firefighters on the scene attempted a rescue but were driven back by intense heat and flames and finally by the collapse of the house's roof. Firefighter Gouckenour's body was found approximately 10 feet inside the front door of the structure. The cause of death was asphyxiation due to smoke inhalation and carbon monoxide. Firefighter Gouckenour joined the fire department after his home burned 2 years before his death. Additional information about this incident can be found in NIOSH Fire Fighter Fatality Investigation 99-F-02.

1/10/99
Tracy Dolan Toomey, Firefighter
Age 52, Career
Oakland Fire Department, California

Firefighter Toomey was crushed and killed when the second floor of a turn-of-the-century residential structure collapsed into the first floor. The fire eventually went to 6 alarms. A total of 4 firefighters were trapped by the collapse, including Firefighter Toomey. Additional information about this incident can be found in NIOSH Fire Fighter Fatality Investigation 99-F-03.

1/11/99
Martin Richard Wauson, Forestry Technician
Age 55, Wildland Part-Time
United States Department of Agriculture Forest Service, Arkansas

Forestry Technician Wauson was participating in an annual requalification test for seasonal firefighting duty with the U.S. Forest Service (USFS). The test required the participant to carry a 45 pound backpack and cover 3 miles within 45 minutes. Forestry Technician Wauson had just passed the 1-mile mark when he collapsed. CPR and defibrillation were begun immediately by medical personnel at the site; however, Forestry Technician Wauson was not revived. The autopsy revealed that Forestry Technician Wauson's death was caused by hypertensive arteriosclerotic disease. As a result of Wauson's death, the USFS temporarily suspended use of the pack test.

1/16/99
James R. Tolan, Firefighter
Age 55, Career
Dunmore Fire Department, Pennsylvania

Firefighter Tolan had responded to a report of a structural fire as the driver and sole occupant of a ladder truck. When he arrived on the scene, he exited his apparatus and told his shift commander that he was not feeling well. He was taken to the hospital by ambulance, was treated for a heart attack, and subsequently was discharged from the hospital. Firefighter Tolan never returned to work. He became ill again in June, was readmitted to the hospital, and died on June 9, 1999.

1/19/99
James H. McGroarty, Firefighter/Fire Investigator
Age 43, Career
Syracuse Fire Department, New York

Fire Investigator McGroarty was in the attic of a residential structure that had experienced a fire 5 days before. A private fire investigator and an electrical consultant also were in the attic with Investigator McGroarty. During the course of the investigation, a chimney, which had been supported by the roof prior before the fire, collapsed onto Investigator McGroarty and inflicted severe injuries. The chimney was too heavy for the personnel on the scene to lift and it stayed in place until additional personnel arrived. Medical treatment was initiated immediately after Investigator McGroarty was freed and was continued at the hospital, to no avail. The cause of death on the autopsy was listed as multiple injuries caused by falling debris from a recent fire. Additional information about this incident can be found in NIOSH Fire Fighter Fatality Investigation 99-F-06.

1/27/99
Ralph J. Loyd, Firefighter
Age 90, Volunteer
Greenville Fire Protection District, Illinois

Firefighter Loyd had been a member of the Greenville Fire Protection District and its predecessors for 70 years. Firefighter Loyd drove a fire department equipment truck to the scene of a reported grain elevator fire. As he was standing by the truck shortly after arrival, he collapsed from a heart attack. Medical aid was rendered at the scene. Firefighter Loyd died the next day, January 28, 1999. No autopsy was performed.

1/29/99
Joseph R. "Dick" Murphy, Firefighter
Age 64, Career
Boston Fire Department, Massachusetts

Firefighter Murphy was preparing to back his command vehicle into the station after returning from a report of smoke in an apartment building. Firefighter Murphy did not complain of any illness at the scene. As he prepared to back in, he collapsed on the steering wheel from a heart attack. The chief officer who was riding in the vehicle attempted to shift the vehicle into park but was unsuccessful. The command vehicle, a Chevrolet Suburban, proceeded in reverse and collided with a pumper located in the apparatus bay. Despite the immediate administration of cardiopulmonary resuscitation (CPR) and automatic external defibrillator (AED) shocks by firefighters and the subsequent arrival of advanced life support (ALS) on the scene, Firefighter Murphy was pronounced dead shortly after arriving at the hospital. The autopsy listed the cause of death as occlusive coronary heart disease. Additional information about this incident can be found in NIOSH Fire Fighter Fatality Investigation 99-F-12.

2/9/99
Gerald "Jerry" Myers, Fire Safety Officer
Age 59, Volunteer
Sumpter Fire Department, Oregon

Firefighter Myers was clearing snow from around fire hydrants with a backhoe. A mechanical problem occurred with the backhoe and Firefighter Myers brought the equipment to a repair facility to fix the problem. During the course of the repairs, Firefighter Myers was struck by a part of the equipment and died of a traumatic injury.

2/12/99
Jimmy "Wayne" Kittle, Captain
Age 49, Career
Whitfield County Fire Department, Georgia

Captain Kittle suffered a massive pulmonary embolism while on-duty at the fire station and died.

2/15/99
Brian William Collins, Assistant Fire Chief
Age 35, Volunteer
River Oaks Volunteer Fire Department, Texas

Phillip Wayne Dean, Captain
Age 29, Volunteer
River Oaks Volunteer Fire Department, Texas

Gary Charles Sanders, Firefighter
Age 20, Volunteer
Sansom Park Fire Department, Texas

Chief Collins, Captain Dean, and Firefighter Sanders were members of an attack team working in the interior of a church that was involved in fire when the roof collapsed and trapped the three in the fire area. The fire was set in a shed next to the church and spread into the attic rafters of the church building itself. Firefighters were attacking an attic fire from the interior of the structure and were being ordered to the exterior as the collapse occurred. Firefighters were on the roof at the time of the collapse and 1 firefighter fell into the church but was not seriously injured. The fire was determined to be an arson.

Assistant Chief Collins died of extensive thermal burns. He had a carboxyhemoglobin level of 6.25 percent. Captain Dean died of smoke inhalation with thermal injuries. He had a carboxyhemoglobin level of 52.5 percent. Firefighter Sanders died of smoke inhalation with thermal injuries. He had a carboxyhemoglobin level of 72.5 percent. Brian Collins was a career Lieutenant and Phillip Dean was a career Fire Engineer with the Fort Worth Fire Department.

2/15/99
Terry Lee Myers, Driver/Operator
Age 50, Volunteer
Vigilant Hose Company, Emmitsburg, Maryland

Driver/Operator Myers was working as a pump operator at the scene of a brush fire on the campus of Mount Saint Mary's College in Emmitsburg, Maryland. He had been working for about 45 minutes when he collapsed of a heart attack. Driver/Operator Myers had not complained of any sickness prior to his attack. Emergency medical care was provided by members of his Department, the local ambulance squad, and by paramedics. Despite their efforts, Driver/Operator Myers was pronounced dead at a local hospital. The brush fire was caused by the spread of an unattended fire being used to dispose of cleared brush and trees. Additional information about this incident can be found in NIOSH Fire Fighter Fatality Investigation 99-F-43.

2/16/99
Robert C. Stanmire, Sr., Firefighter
Age 52, Volunteer
Forest Grove Volunteer Fire Company, Vineland, New Jersey

Firefighter Stanmire suffered a heart attack as he and other members of his department prepared to respond to a brush fire. Firefighter Stanmire was stricken in the fire station as he boarded a piece of fire apparatus while dressed in full turnouts. Fellow firefighters began CPR immediately, and ALS was provided by the local rescue squad, but Firefighter Stanmire was dead upon arrival at a local hospital. The brush fire was arson. No autopsy was performed.

2/18/99
Burton Frank Chestnut, Firefighter
Age 67, Volunteer
Brazos Volunteer Fire Department, Texas

Firefighter Chestnut was on a personal errand with two other Brazos firefighters when they came upon a fire in a woodpile that was against a structure. Firefighter Chestnut remained at the scene to begin notifying residents of the fire while other firefighters went to retrieve their fire apparatus. When the other firefighters returned, they found Firefighter Chestnut dead of an apparent heart attack.

2/19/99
Terry "Ted" Oliver, Assistant Fire Chief
Age 58, Volunteer
Eaton Rapids Fire Department, Michigan

Assistant Chief Oliver was directing the overhaul and salvage of a bedroom fire that had been extinguished. He ascended the stairs to the second floor of the house to check on the progress of work in the fire area. Upon reaching the top of the stairwell, he collapsed and died of an apparent heart attack. He was removed from the structure and taken to the hospital by an ambulance that was on the scene of the fire. Assistant Chief Oliver's son is a firefighter and he was on the scene when his father was stricken. He was the first firefighter to die in the line of duty for the 125-year-old department.

2/28/99
Alan W. Ducheck, Captain
Age 46, Paid-on-Call
DeSoto Fire and Rescue, Missouri

Captain Ducheck was assisting with an extended vehicle extrication when he suffered a heart attack. He died on March 1, 1999.

3/12/99
Jerome Taylor, Captain
Age 69, Volunteer
Hillburn Fire Department, New York

Captain Taylor collapsed and died of an apparent heart attack while directing traffic at a structure fire.

3/16/99
David L. Packard, Firefighter
Age 56, Career
Boston Fire Department, Massachusetts

Firefighter Packard was preparing for duty when his engine company was dispatched to a motor vehicle collision. Firefighter Packard did not respond on the call (he was on the oncoming shift and the call was handled by the off going shift) and was found by other firefighters upon their return to quarters from the call. He had collapsed in the bunkroom of an apparent heart attack. Despite immediate aid by on-scene firefighters, the use of an AED, and ALS treatment by paramedics, Firefighter Packard died. No autopsy was performed. The cause of death as listed on his death certificate was asystole caused by coronary artery disease. Additional information about this incident can be found in NIOSH Fire Fighter Fatality Investigation 99-F-17.

3/16/99
Charles James "Chuck" Vodak, Firefighter
Age 45, Volunteer
Dunning Fire Department, Nebraska

Firefighter Vodak and other members of his department had been fighting a prairie fire for more than 4 hours. Firefighter Vodak complained of chest pains and experienced a heart attack. Other firefighters performed CPR on Firefighter Vodak for 1 hour as they waited for an ambulance to arrive at the isolated fire location. The path of the fire was 10 miles wide in some places and it eventually consumed 70,000 acres.

3/17/99
Walter J. Flyntz, Firefighter
Age 44, Career
Bridgeport Fire Department, Connecticut

Firefighter Flyntz responded as part of an engine company to a report of a fire in the basement of an apartment building that was undergoing renovation. Upon arrival at the scene, he assisted with the connection of his engine to a hydrant and then helped check for fire extension on the upper floors of the building. After clearing the second floor, Firefighter Flyntz went to the third floor to check for extension. Shortly after his arrival on the third floor, he collapsed. He was discovered by a renovation worker who, in turn, notified other firefighters. Despite the efforts of firefighters on the scene and efforts by other emergency workers, Firefighter Flyntz expired. The cause of death was listed as atherosclerotic cardiovascular disease. Additional information about this incident can be found in NIOSH Fire Fighter Fatality Investigation 99-F-18.

3/23/99
Paul Haislopp, Captain
Age 50, Volunteer
Centerville Fire Department, Ohio

Captain Haislopp experienced a heart attack at his fire station as he moved fire apparatus in preparation for a response to a structure fire.

4/2/99
Aubrey R. Tillman, Firefighter
Age 57, Career
Charleston Fire Department, South Carolina

Firefighter Tillman had been on-duty for 14 hours and had participated in physically demanding training for 3 ½ hours of his shift. At approximately 10 p.m., Firefighter Tillman experienced a witnessed temporary loss of consciousness. As other firefighters came to his aid, Firefighter Tillman regained consciousness and complained of severe chest pains. He was transported by ambulance to the hospital under ALS care. His condition deteriorated in the ambulance and further deteriorated in the hospital emergency room. Firefighter Tillman died at 11:37 p.m. The death certificate listed "probable acute myocardial infarction" as the immediate cause of death. No autopsy was performed. Additional information about this incident can be found in NIOSH Fire Fighter Fatality Investigation 99-F-15.

4/6/99
Kenneth Allen Nickell, Captain
Age 28, Volunteer
Route 377 Volunteer Fire Department, Kentucky

Kevin Rex Smith, Firefighter/EMT
Age 30, Volunteer
Route 377 Volunteer Fire Department, Kentucky

Captain Nickell and Firefighter Smith responded to a wildland fire in the Daniel Boone National Forest near Cranston, Kentucky. They were part of a 7-person team that was constructing a fire line in hardwood leaf litter on the forest floor. Nickell and Smith were in the lead and using a rake and a gasoline-powered leaf blower to construct the line. As the fire line was being constructed, spot fires were breaking over the fire line and several members of the team doubled back to control the spot fires. Captain Nickell and Firefighter Smith continued to construct the fire line. The fire was growing in intensity and the wind was picking up so the crew leader gave the order for all firefighters to pull back. Captain Nickell acknowledged the order and indicated that he and Firefighter Smith would pull back. Shortly thereafter, another radio transmission was received from Captain Nickell indicating that he and Firefighter Smith were burned or on fire. Evidence suggests that the two tried to outrun the fire uphill but were slowed by terrain. It appeared as if the firefighters attempted to run back through the fire to reach the burned area. At some point, they succumbed to the flames and collapsed. The cause of death for both firefighters was listed as asphyxia due to environmental oxygen deprivation, smoke inhalation, and acute carbon monoxide poisoning. Neither firefighter was equipped with a fire shelter. Additional information about this incident can be found in NIOSH Fire Fighter Fatality Investigation 99-F-14. A report also is available from the Kentucky Division of Forestry entitled "Report of the Accident Investigations Team for the Island Fork Fire – April 6, 1999 – Near Cranston, Kentucky."

4/8/99
John E. Murphy, Deputy Fire Chief
Age 64, Paid-on-Call
Russell Fire Department, Massachusetts

Deputy Chief Murphy was engaged in fighting a wildland fire for 4 hours. He collapsed and was down for 10 to 15 minutes before being discovered by another firefighter. CPR was begun immediately thereafter and continued as Deputy Chief Murphy was transported to a waiting ambulance in the bed of the Fire Chief's pickup (due to terrain). CPR continued through transport in the BLS ambulance, ALS treatment, and treatment in a hospital emergency room. Despite all efforts, Chief Murphy expired in the emergency room. The cause of death was listed as coronary atherosclerosis. Additional information about this incident can be found in NIOSH Fire Fighter Fatality Investigation 99-F-32.

4/12/99
Phillip M. Pinkowski, Jr., Firefighter
Age 59, Volunteer
Clarendon Volunteer Fire Department, Vermont

Firefighter Pinkowski collapsed and died of an apparent heart attack while acting as a pump operator at a residential structure fire.

4/15/99
Robert D. Peters, Firefighter
Age 71, Volunteer
West Lake Fire Department, Pennsylvania

Firefighter Peters responded to an ambulance call and was not feeling well at the call. After the call was concluded, he went home. That afternoon, he was found at home by a relative. He was nearly unconscious. Despite surgery, Firefighter Peters died as the result of a stomach aneurysm.

4/28/99
David J. Watts, Captain
Age 56, Paid-on-Call
Nantucket Fire Department, Massachusetts

Captain Watts was engaged in active structural firefighting for 2 ½ hours in a multiple-occupancy wood frame building that was constructed about 1849. After the fire was controlled, Captain Watts returned home and suffered a heart attack in the shower shortly after arriving home. CPR was administered, first by his wife and later by firefighters, and an AED was used by other firefighters who responded to Captain Watts' residence. He was transported by ambulance to a local hospital where he was stabilized. Captain Watts was transferred to a hospital in Boston by air but he never regained consciousness and died on May 2, 1999. The cause of death was listed as coronary thrombosis and ventricular tachycardia. Additional information about this incident can be found in NIOSH Fire Fighter Fatality Investigation 99-F-19.

5/2/99
Kenneth Alan Strain, Firefighter
Age 28, Volunteer
Hemby Bridge Volunteer Fire Department, North Carolina

Firefighter Strain was the sole occupant and driver of a 1996 pumper. He was returning to the station after being canceled en route to a motor vehicle collision. He pulled the pumper to the right side of the road to allow for the passage of other traffic when the right rear wheels of the pumper left the paved surface of the road and fell into a ditch. Firefighter Strain was unable to regain control, and the pumper struck a tree. Firefighter Strain was killed instantly and was pronounced dead at the scene. According to the highway patrol, speed was not a factor. Firefighter Strain was wearing his seatbelt. It took other firefighters 2 hours to free Firefighter Strain from the wreckage. Additional information about this incident can be found in NIOSH Fire Fighter Fatality Investigation 99-F-16.

5/3/99
Eric Noel Casiano, Firefighter
Age 41, Career
Philadelphia Fire Department, Pennsylvania

Firefighter Casiano and his company were fighting a structural fire in a residential occupancy. During the fire fight, Firefighter Casiano fell through a floor but appeared to be uninjured. After his company had been released and was back at quarters, everyone returned to bed. A short time later, Firefighter Casiano's company was dispatched to another call. Other firefighters found him down and without vital signs. Despite immediate CPR and ALS arrival within 4 minutes, Firefighter Casiano died. His autopsy revealed that he had a massive hemorrhage within his back muscles which damaged his spinal cord. The hemorrhage was caused by the fall.

5/4/99
Arthur A. Tullis, Fire Chief
Age 61, Part-Time (Paid)
LaGrange Park Volunteer Fire Department, Illinois

Chief Tullis and members of his fire department and a neighboring department had responded to an automatic fire alarm activation in a retirement home. Chief Tullis was first on the scene and was exiting the building to command the arrival of other units when he collapsed from an apparent heart attack. ALS was provided immediately by other firefighters, but Chief Tullis could not be revived. The cause of death was heart disease. The alarm activation was caused by carpet installers.

5/14/99
Lewis Edward Williams, Fire and Rescue Captain
Age 47, Volunteer
Fort Oglethorpe Fire & Rescue, Georgia

Captain Williams worked for 2 ½ hours on the scene of a trench collapse which trapped one worker. He participated in various tasks, including the unloading of supplies, and command. He collapsed at the command post without warning or any complaint of sickness. Despite immediate EMS and ALS care, Captain Williams died. The cause of death was listed as a heart attack. The trapped construction worker was rescued successfully. Additional information about this incident can be found in NIOSH Fire Fighter Fatality Investigation 99-F-49.

5/30/99

Lewis Jefferson Matthews, Firefighter	**Anthony Sean Phillips, Sr., Firefighter**
Age 29, Career	Age 30, Career
District of Columbia Fire Department	District of Columbia Fire Department

Firefighter Matthews and Firefighter Phillips were members of two different engine companies working on the first floor of a townhouse that was experiencing a fire. Both crews had entered the front door of the townhouse at street level. The fire was confined to the basement. The basement, at grade at the rear of the structure, was opened by a truck company and a small fire was observed. A company officer at the basement door requested permission to hit the fire but his request was denied by the incident commander since he knew that crews were in the building and he did not want to have an opposing hose stream situation. The fire grew rapidly and extended up the basement stairs into the living areas of the townhouse where Firefighter Matthews, Firefighter Phillips, and other firefighters were working.

With the exception of Firefighter Matthews and Firefighter Phillips, all firefighters exited the building after the progress of the fire made the living area of the townhouse untenable. On the exterior of the building, firefighters realized that Firefighter Matthews was not accounted for. Firefighters reentered the building and followed the sound of a PASS device. They removed the firefighter with the activated PASS to the exterior of the building. Once outside, firefighters realized that the firefighter who had been rescued was not Firefighter Matthews, but was, in fact, Firefighter Phillips. The search continued and Firefighter Matthews was discovered and removed approximately 4 minutes later.

Firefighter Phillips' PASS device was of the type that is automatically activated when the SCBA is activated and it worked properly. Firefighter Matthews' PASS was a manually activated type and it was found in the "off" position.

Both firefighters received immediate medical care on the scene and were transported rapidly to hospitals. Firefighter Phillips was pronounced dead upon arrival at the hospital and Firefighter Matthews died the following day, May 31, 1999.

Firefighter Phillips died as the result of burns over 60 percent of his body surface area and his airway. Firefighter Matthews died as the result of burns over 90 percent of his body surface area and his airway. 2 other firefighters were injured fighting the fire. One (1) of these 2 firefighters, who suffered burns over 60 percent of his body surface area, survived and was released from the hospital in late August. At the time of his release, it was not clear if this firefighter would ever return to work.

Additional information about this incident can be found in NIOSH Fire Fighter Fatality Investigation 99-F-21.

5/31/99
Joseph F. Tagliareni, Jr., Firefighter
Age 34, Volunteer
Secaucus Fire Department, New Jersey

Firefighter Tagliareni was driving an engine company in response to a report of an automobile fire. He started to feel ill, pulled the apparatus to the side of the road, exited the apparatus, walked to the rear of the engine, and collapsed of a heart attack. Other firefighters began CPR immediately and ALS-level care was provided before transporting Firefighter Tagliareni to the hospital. Firefighter Tagliareni was in a coma for 13 days before his death on June 13, 1999. The autopsy found hypertrophy and dilatation of the heart, and pneumonia.

6/2/99
Rudolf Cohen, Deputy Fire Chief
Age 67, Career
Gary Fire Department, Indiana

Chief Cohen collapsed and died of heart failure while working at his desk. No autopsy was performed.

6/3/99
Vincent Fowler, Captain
Age 47, Career
Fire Department City of New York, New York

Captain Fowler and a probationary firefighter were searching for victims and fire in the basement of an occupied residential structure. The basement was extremely congested with furniture, newspapers, magazines, and other items. Water had been applied to the fire and it appeared to be under control. The low air alarms on Captain Fowler's SCBA and the probationary firefighter's SCBA were sounding. Captain Fowler delayed his own exit from the basement as he called for and accounted for his probationary firefighter and he subsequently ran out of air. Captain Fowler and the probationary firefighter buddy-breathed until Captain Fowler collapsed. The probationary firefighter dragged Captain Fowler until he was joined by other firefighters. Because of the congested conditions in the basement, it took an extended period of time for Captain Fowler to be removed from the building. Captain Fowler died the next day on June 4, 1999.

The cause of death was listed as smoke inhalation with carbon monoxide inhalation. Captain Fowler's carbon monoxide level was 60 percent.

6/4/99
Richard Anthony Heinze, Firefighter
Age 47, Career
Newark Fire Department, New Jersey

Firefighter Heinze had just returned from a fire run that turned out to be a malfunctioning fire alarm system. He retired to the bunkroom shortly after returning to the station. Approximately 40 to 50 minutes after returning to the station, other firefighters heard a loud "thump" and found Firefighter Heinze on the floor of the bunkroom gasping for air. He soon became unconscious. Firefighters immediately began CPR and an ambulance was summoned. Despite medical care that included defibrillation, Firefighter Heinze was pronounced dead at the hospital. Firefighter Heinze had complained of pain in his jaw and a headache earlier in the shift. The cause of death was listed as hypertensive and atherosclerotic cardiovascular disease. Additional information about this incident can be found in NIOSH Fire Fighter Fatality Investigation 99-F-31.

6/8/99
Clyde Peterson, Assistant Fire Chief
Age 70, Volunteer
Hilltop Lakes Volunteer Fire Department, Texas

Chief Peterson had just finished commanding the response of his fire department to a fire involving a stake bed truck filled with asphalt shingles. As he was returning to his vehicle, he was struck by an apparent heart attack. He was discovered by other firefighters when they noticed that he was missing. CPR was begun immediately. No autopsy was performed. The cause of the fire was listed as suspicious.

6/9/99
Phillip P. Cirrito, Houseman/Firefighter
Age 47, Career
Merion Fire Company of Ardmore, Pennsylvania

Firefighter Cirrito had been on-duty for approximately 22 hours. During the shift, he had responded to 2 fire incidents and had assisted with the installation of radio equipment on a new pumper. The day was extremely hot with temperatures above 90° F. At approximately 5:50 a.m., an ambulance was dispatched to the fire station to a report of a firefighter gasping for air. Firefighter Cirrito died sometime later that day. The cause of death was listed as a pulmonary thromboembolism and hypertensive cardiovascular disease.

6/13/99
Arch Russell Sligar, Jr., Firefighter
Age 52, Volunteer
Bethany Volunteer Fire Department, West Virginia

Firefighter Sligar was riding in the officer's seat of a pumper responding to a mutual-aid structure fire. The apparatus driver noticed that Firefighter Sligar was not talking and that he had slumped over in the officer's seat. The apparatus was stopped, Firefighter Sligar was removed from the apparatus, and CPR was begun. ALS arrived within 2 minutes and defibrillated Firefighter Sligar. He regained consciousness for a short time but suffered another arrest in the ambulance en route to the hospital. He was pronounced dead about 40 minutes after arrival at the hospital. The cause of death was listed as a heart attack. No autopsy was performed. Additional information about this incident can be found in NIOSH Fire Fighter Fatality Investigation 99-F-24.

6/11/99

Wayne Robert Luecht, Assistant Chief

Age 47, Career

Northbrook Fire Department, Illinois

Firefighters from the Northbrook Fire Department responded to a report of an electrical problem in a department store located in a mall. Crews discovered a smell of smoke in the building and located some electrical equipment that had burned. There was no hazard of fire extension so crews were released to return to quarters and the on-duty Fire Prevention staff were requested to the scene. Assistant Chief Luecht, the Fire Marshal, arrived and accompanied a private electrical contractor as he investigated the cause of the power outage and fire. As the electrician was testing an electrical panel, a white-blue flash, concussion, and fireball occurred, enveloping the electrician and Chief Luecht. Although severely burned over 90 percent of his body surface, Chief Luecht directed the response to the emergency and directed firefighters to assist other victims before he allowed them to provide him treatment. Chief Luecht was alert and oriented throughout his treatment and spoke with others until he was placed on a ventilator at the hospital. Chief Luecht fought for 10 days until he succumbed to his injuries on June 21, 1999. The electrician was killed and 2 department store employees were injured.

6/16/99

Clifford Thomas Moore, Fire Captain

Age 38, Career

Manteca Fire Department, California

Captain Moore was participating in training exercises at a regional training facility. Captain Moore was proctoring the hose aloft and ladder rescue evolutions. During the second evolution, the incident commander initiated an emergency evacuation of the building as part of the training. Captain Moore attempted an emergency evacuation method from the second story window of the training tower onto a ground ladder that had been placed at the window. Captain Moore fell from the window and received critical facial and head injuries. He was treated by paramedics on the scene and transported to the hospital. Despite aggressive efforts to save his life, Captain Moore was pronounced dead at the hospital. The Manteca Fire Department conducted a board of inquiry into the incident. The report is available for download at the Manteca Fire Department Web site. Additional information about this incident can be found in NIOSH Fire Fighter Fatality Investigation 99-F-25.

6/17/99
Paul Francis McGrath, Firefighter
Age 50, Career
Pittsburgh Fire Department, Pennsylvania

Firefighter McGrath was a member of a truck company that was fighting a 3 alarm defensive fire in a 3-story brick building that had last been used as living quarters for nursing students. Firefighter McGrath participated in numerous tasks on the fireground, including establishing a water supply to his truck company for master stream operations, ventilation, placement of ground ladders, and forcible entry. The fire had just been brought under control when the call went out that a firefighter was down. Firefighter McGrath had become dizzy at the aerial ladder turntable while operating the ladder pipe and was assisted to the ground by other firefighters. Upon reaching the ground, he collapsed of an apparent heart attack. ALS was provided by on scene EMS crews and Firefighter McGrath was transported to the hospital but later died. The cause of death was listed as "arteriosclerotic cardiovascular disease." The fire was found to be arson, and suspects were arrested. Firefighter McGrath was born at the hospital located on the same grounds as the nurse's residence that burned. Additional information about this incident can be found in NIOSH Fire Fighter Fatality Investigation 99-F-22.

6/18/99
Ronald Gregory Phillips, Firefighter Paramedic
Age 32, Career
Sylvania Township Fire Department, Ohio

Firefighter Phillips had just begun his daily physical fitness workout when he collapsed. He had been working out with dumbbells on a bench. His collapse was not witnessed but others in the area heard noise and found him unconscious. ALS treatment was provided and he was transported to the hospital. Despite every effort, Firefighter Phillips never regained consciousness and died. An autopsy was performed and the cause of death was listed as a ruptured cerebral aneurysm.

6/23/99

Matthew Eric Black, Firefighter

Age 20, Volunteer

Lakeport Fire Department, California

A large branch from a mature oak tree fell on some power lines and brought them down. The Lakeport Fire Department was dispatched to control a grass fire that resulted from the fallen power lines. Firefighters were advised upon dispatch that power lines were down. Firefighter Black's place of work was about 1½ miles from the fire scene, so he responded directly to the scene in his personal vehicle and joined up with an engine company. Firefighter Black asked if he could advance a booster reel line and extinguish a pile of burning debris; and permission was granted. According to witness accounts, Firefighter Black appeared to stumble after heaving on the hose after it hung up. Firefighter Black fell face down on the live wire and was electrocuted. Other firefighters on the scene used a handtool to remove the wire from under Firefighter Black and dragged him away from the wire. He was found to have no pulse or respiration. CPR was begun immediately. ALS arrived within 9 minutes and Firefighter Black was transported to the hospital. Despite aggressive efforts to revive him, Firefighter Black was pronounced dead at the hospital. The cause of death was listed as electrocution. Additional information about this incident can be found in NIOSH Fire Fighter Fatality Investigation 99-F-26.

7/4/99

Roger B. McEwen, Sr., Captain

Age 52, Volunteer

Hanover Volunteer Fire Department, Alabama

Captain McEwen had been working on the scene of a mobile home fire for about 1 hour. During the course of the fire fight, Captain McEwen had advanced and operated hose lines, had helped other firefighters with their SCBA's, and was operating a fire pump at the time of his illness. He cried out for help and collapsed to the ground, the victim of an apparent heart attack. ALS providers were on scene and began care immediately. Despite efforts on the scene and during transport, Captain McEwen was pronounced dead at the hospital. No autopsy was performed.

7/7/99
Costello Nathaniel "Colonel" Robinson, Firefighter/Technician
Age 64, Career
District of Columbia Fire Department

Firefighter Robinson was the most senior active firefighter on the District of Columbia Fire Department. He and his engine company were dispatched, along with other units, to the report of a fire in a densely populated area of the District. As Firefighter Robinson and other firefighters were searching for a fire, Firefighter Robinson was attacked from behind by an unrestrained pit bull terrier dog. Firefighter Robinson was injured by the attack and unable to walk. He was transported to a local hospital and scheduled for knee surgery to repair the damage caused by the dog attack. On July 9th, the day that his knee surgery was scheduled, Firefighter Robinson became acutely short of breath and unresponsive. Medical aid was provided, but Firefighter Robinson did not survive. Autopsy findings included "hypertensive cardiovascular disease" and "blunt impact trauma [to the knee] with avulsion of [the] right quadriceps tendon." Additional information about this incident can be found in NIOSH Fire Fighter Fatality Investigation 99-F-40.

7/7/99
Lawrence D. Lehman, Fire Police Lieutenant
Age 51, Volunteer
South Lebanon Township Fire Police - Friendship Fire Company, Pennsylvania

Fire Police Lieutenant Lehman was performing crowd control and traffic duties at the scene of a wildland fire. Firefighters had been on the scene for less than 1 hour when Fire Police Lieutenant Lehman collapsed of an apparent heart attack. EMS was provided immediately; however, Fire Police Lieutenant Lehman was pronounced dead at the hospital.

7/12/99
David Vernon Parks, Firefighter Photographer
Age 69, Volunteer
Washington Township Volunteer Fire Department, Pennsylvania

Firefighter Parks was engaged in routine maintenance of an engine company apparatus when he was stricken with a heart attack and later died.

7/15/99
Bryan Christopher Pottberg, Firefighter Paramedic
Age 25, Career
Lee's Summit Fire Department, Missouri

Firefighter Paramedic Pottberg was participating in a scheduled rescue diver drill. Firefighter Paramedic Pottberg was underwater performing a search evolution when he failed to surface. Other firefighters searched for him. He was brought to the surface approximately 11 minutes after firefighters noticed that something was wrong. He received immediate medical attention in the boat and while en route to the hospital. He was pronounced dead upon arrival at the hospital. The autopsy listed the cause of death as drowning. It is unknown why Firefighter Paramedic Pottberg had difficulty. Additional information about this incident can be found in NIOSH Fire Fighter Fatality Investigation 99-F-29.

7/18/99
Martin Michael Stiles, Inmate Firefighter
Age 40, Wildland Part-Time
Los Angeles County Fire Department, California

Firefighter Stiles was a part of a Los Angeles County Fire Department Strike Team working a wildland fire incident in Ventura County. His crew was assigned to construct a handline around a slopover that extended over a dozer line on a ridge. Firefighter Stiles slipped over a ridge and fell to his death 150 feet below. The time of the fatal incident was approximately 1:50 a.m.

7/27/99
David Dwayne Hartwick, Firefighter
Age 35, Volunteer
New Braunfels Fire & Rescue, Texas

Firefighter Hartwick was at home asleep when his fire department was dispatched to a structure fire. Firefighter Hartwick was found dead in the morning, the victim of a heart attack. His pager had been reset after the page for the structure fire. It is assumed that Firefighter Hartwick rose to respond to the structure fire and then died of the heart attack. This fact cannot be established with 100 percent certainty.

7/29/99
Richard F. Devine, Firefighter
Age 49, Career
Philadelphia Fire Department, Pennsylvania

Firefighter Devine was assigned to an engine company working on the scene of a structure fire. Firefighter Devine was the nozzle person and extinguished a fire on the second floor of the structure. After taking a short break outside of the structure after the fire was controlled, Firefighter Devine re entered the structure to assist with overhaul. Upon his arrival at the second floor, he collapsed of an apparent heart attack. CPR was begun immediately by other firefighters, ALS was provided by fire department paramedics, and Firefighter Devine was transported to the hospital. Firefighter Devine was pronounced dead at the hospital when further medical efforts proved fruitless. The autopsy listed the cause of death as "arteriosclerotic cardiovascular disease" with heat stress cited as another significant condition. The fire was caused by a child playing with matches. Additional information about this incident can be found in NIOSH Fire Fighter Fatality Investigation 99-F-50.

7/30/99
Kenneth F. Clinch, Firefighter
Age 52, Volunteer
Mount Marion Fire Department, New York

Firefighter Clinch and members of his fire department were fighting a car fire on the New York State Thruway. Firefighter Clinch had driven a water tanker (tender) to the scene and was directed to stretch a 2-1/2" line from the tanker to the pumper. The day was hot and humid. He stretched 100 feet of line from the tanker to the pumper, made the connection to the pumper, and was headed back to the tanker when he collapsed of an apparent heart attack. On scene firefighters immediately went to his aid and ALS was requested. The ambulance arrived within 8 minutes. Firefighter Clinch was breathing and had a weak pulse when he left the scene for the hospital; however, he died later that day. The autopsy reported the cause of death as occlusive coronary artery disease.

8/5/99
Richard Clarence Bacon, Assistant Chief
Age 46, Volunteer
Dunstable Volunteer Fire Department, Massachusetts

Assistant Chief Bacon had driven a fire department pumper to the scene of a reported structure fire. The fire was minor, and all units were ordered to return to the station. Upon returning to the station, Assistant Chief Bacon was completing some paperwork when he was observed to be in medical distress. Firefighter/EMT's in the station began treatment, including CPR, and Chief Bacon was transported to the hospital. Assistant Chief Bacon did not survive his heart attack.

8/5/99
James Everett. Clark, III, Senior Firefighter
Age 42, Career
Midwest City Fire Department, Oklahoma

Senior Firefighter Clark was a member of a squad company that had been dispatched to the report of a motor vehicle collision on I-40 in Midwest City. The roads were wet from rain and rain had begun to fall again. A ladder company also was dispatched on the call. The squad arrived on the scene and discovered that the collision was minor. The ladder arrived and positioned itself upstream of the squad to divert traffic away from the incident scene; all of the unit's emergency lights were operating. Approximately 2 minutes after arriving on the scene, the ladder was hit from behind by a passenger vehicle. Firefighters dismounted the ladder apparatus to check on the condition of the driver. The ladder firefighters were joined by Senior Firefighter Clark, who had heard the collision. After the patient from the second collision was moved to an area that was thought to be safe (between the ladder apparatus and the wall), the company officer of the ladder company walked further upstream in traffic in an attempt to wave traffic away from the scene of both collisions. At this point, another passenger vehicle lost control and spun into the space between the ladder apparatus and the retaining wall. Senior Firefighter Clark placed himself between the oncoming car and the driver of the car that had collided with the ladder apparatus. 2 firefighters and the driver of the car that hit the ladder apparatus were injured. The injured received immediate medical care on the scene, en route to the hospital, and after arriving at the hospital. Senior Firefighter Clark died as the result of head injuries on August 8, 1999. Additional information about this incident can be found in NIOSH Fire Fighter Fatality Investigation 99-F-27.

8/5/99
Michael Eugene "Cuppie" Cupp, Sr., Fire Chief
Age 48, Volunteer
Cygnet Volunteer Fire Department, Ohio

Chief Cupp and members of his fire department had just completed the extinguishment of a brush fire on a vacant lot. As Chief Cupp returned to his truck, he collapsed of an apparent heart attack. Despite EMS care on the scene, en route to the hospital, and in the hospital, Chief Cupp was pronounced dead later that night. The cause of death was a massive myocardial infarction. No autopsy was performed.

8/5/99
Cilton Jay Dauzat, Firefighter
Age 63, Volunteer
White Tail Ridge Volunteer Fire Department, Texas

Firefighter Dauzat had responded with members of his fire department to a wildland fire that resulted from the failure of a local homeowner to contain an intentional fire in a pile of logs. Unbeknownst to the driver of a pumper, Firefighter Dauzat had mounted the back step. The pumper was attempting to ascend a hill. Despite two tries, the pumper was unable to climb the hill and when the apparatus backed down off of the hill, the driver discovered the he had run over Firefighter Dauzat. It is believed that Firefighter Dauzat lost his footing. The cause of death was listed as severe chest and head injuries.

8/8/99
Ronald Wade Meshell, Firefighter
Age 30, Volunteer
Huttig Fire Department, Arkansas

Firefighter Meshell responded, along with others from his department, to a mutual-aid call for a fire in a motor home. The owner of the motor home had just filled the fuel tank, driven the motor home to his residence, and parked the vehicle at the top of his driveway. Upon arriving home, the owner noticed smoke coming from under the hood; he could not extinguish the fire and called the fire department. Firefighter Meshell and a Deputy Fire Chief arrived in a pumper. The Huttig pumper arrived on the scene and parked about 40 feet downhill from the motor home. Firefighter Meshell was ordered to pull a 1-1/2" line from the rear of the Huttig pumper to assist a firefighter from another department that had a booster line from the other department's pumper on the fire. Before the line could be charged, the fuel tank on the motor home ruptured and sent a flood of burning fuel down hill toward firefighter Meshell, another firefighter, and both pumpers. Firefighter Meshell was not wearing any firefighting protective clothing, although he had loaded his protective clothing on the pumper prior to response. He was surrounded by flames for an estimated 15 seconds and was burned over 96 percent of his body surface and his airway. Firefighter Meshell died on August 16, 1999. Additional information about this incident can be found in NIOSH Fire Fighter Fatality Investigation 99-F-34.

8/9/99

Arthur J. Heckman, Firefighter

Age 66, Volunteer

Macks Creek Fire Department, Missouri

Firefighter Heckman was the first to arrive at a wildland fire that was threatening structures. Firefighter Heckman had pulled a pumper behind a house to fill it with water. He was found slumped over the wheel, the victim of an apparent heart attack.

8/14/99

Frank William Wood, Firefighter

Age 54, Volunteer

Flourtown Fire Company, Pennsylvania

Firefighter Wood and members of his department were fighting a room fire in a nursing home. Firefighter Wood assisted with the rescue of an unconscious staff member and then collapsed of an apparent heart attack. He was rushed to the hospital but subsequently died.

8/26/99

David Thomas Nall, Assistant Chief

Age 40, Volunteer

Town of Jay Volunteer Fire Department, Florida

Assistant Chief Nall responded to an emergency medical incident on a high school football field. Chief Nall assisted with the treatment of the patient and was helping to load the child in an ambulance when he complained of chest pains. He collapsed and CPR was initiated. Chief Nall was transported to a local hospital where he was pronounced dead about 1 hour later. No autopsy was performed.

8/27/99

Stephen Joseph Masto, Firefighter

Age 28, Firefighter

Santa Barbara Fire Department, California

Firefighter Masto was working as an EMT at a wildland fire, roaming among other firefighters and providing first aid to anyone who became injured. He worked a 6 a.m. to 6 p.m. shift, was equipped with a portable radio, and carried a canteen. He did not return to camp at the end of his shift, and a search was initiated. Firefighter Masto was discovered dead about 12 hours later in steep terrain. The cause of death was found to be heat stroke; there was no evidence of trauma or other medical conditions that contributed to his death. Firefighter Masto's shift as an EMT was on August 27, 1999 and he was found on August 28, 1999. Firefighter Masto was wearing brush gear.

8/31/99
Timmy Roger Dawson, Firefighter
Age 34, Volunteer
Center Rock Volunteer Fire Department, South Carolina

Firefighter Dawson was the driver of a 1994 pumper responding to a motor vehicle collision. Two other firefighters were passengers in the pumper. The right wheels of the pumper left the road and Firefighter Dawson attempted unsuccessfully to bring the truck back under control. Firefighter Dawson overcompensated, and the pumper went off the left side of the road, through a yard, and rolled several times. The pumper's speed was estimated at 60 miles per hour in a 35 mile per hour zone. Firefighter Dawson was not wearing a seatbelt. The other firefighters riding on the pumper received minor injuries. The cause of death was listed as blunt trauma.

9/13/99
Kenneth C. Cashman, Firefighter
Age 29, Volunteer
Auglaize Township Volunteer Fire Department, Ohio

Firefighter Cashman was responding to a stove fire in a residence. A dump truck loaded with stone pulled out in front of Firefighter Cashman and his car collided with the dump truck. Firefighter Cashman was killed instantly. Firefighter Cashman was wearing a seatbelt at the time of the collision and was operating a red dash light and a siren. The dump truck driver was charged with aggravated vehicular homicide.

9/15/99
Terri LeAnn Hood, Firefighter
Age 31, Volunteer
McColloch Volunteer Fire Department, Indiana

Firefighter Hood was helping firefighters from a 5-county area battle a 450-acre wildland fire. Firefighter Hood was protecting a tobacco barn with other firefighters when conditions worsened and firefighters decided to withdraw. In heavy smoke and confusing conditions, Firefighter Hood was run over by a pumper backing away from the barn. She was killed instantly. Additional information about this incident can be found in NIOSH Fire Fighter Fatality Investigation 99-F-35.

9/27/99
Lewis Edward "Rawhide" Anderson, Firefighter
Age 68, Volunteer
River Falls Volunteer Fire Department, South Carolina

Firefighter Anderson was directing traffic around an earlier traffic collision. He was positioned about a 1/2 mile north of the collision scene. The weather was rainy and Firefighter Anderson was wearing bright yellow rain gear and using a stop/slow sign. He was struck by an 18-wheel truck which also struck a fire department vehicle on the scene. The truck then left the scene. Firefighters were notified by another truck driver that Firefighter Anderson was down. They went to his aid and administered CPR and other medical care. Firefighter Anderson died on September 30, 1999. According to the police report, the truck driver was operating his vehicle at a reckless speed. According to the certificate of death, Firefighter Anderson was killed by a cerebral contusion and edema due to blunt force trauma of [the] head.

10/3/99
Gregory Edwin Pacheco, Firefighter
Age 20, Wildland Part-Time
Carson National Forest, United States Forest Service, New Mexico

Firefighter Pacheco was a member of a forest firefighting crew constructing a fire line on the La Jolla fire near San Diego, California. He was ascending steep terrain when a large rock fell and hit Firefighter Pacheco on the head, injuring him severely. One (1) other firefighter received moderate injuries and was released back to his crew. Firefighter Pacheco died on October 5, 1999. The cause of death was listed as a closed head injury.

10/4/99
Jeffrey Scott Thompson, Firefighter
Age 20, Volunteer
Howell County Rural Fire District #1, Missouri

Firefighter Thompson and other members of his fire department responded to a grass fire that was caused when some powerlines fell. The line had fallen on a fence, energizing it. The Fire Chief warned all firefighters that fences in the area were energized and to avoid them. Firefighter Thompson and 2 other firefighters were advancing a booster line to control the fire when they came in contact with a fence. All 3 were electrocuted. EMS care was provided by other firefighters on the scene and the injured were evacuated by air. Firefighter Thompson died that day. The 2 other firefighters survived their injuries, although some of the injuries were very serious.

10/5/99

William Malcolm Bethune, Captain
Age 58, Career
Texas City Fire Department, Texas

Captain Bethune was riding in the officer's seat of an engine company responding, with lights and siren, to a medical emergency. As the engine company entered an intersection against the red light, it struck a passenger car, veered off the roadway, and struck a cement freeway support column. Captain Bethune, who was not wearing a seatbelt, was ejected through the front windshield of the pumper. He struck the pavement and received severe injuries. Captain Bethune was provided with EMS care on the scene and flown to a trauma center. He was pronounced dead upon arrival at the trauma center. The cause of death was listed as blunt trauma to the head. The driver of the pumper also was severely injured; the firefighter riding in the back of the cab had only minor injuries. A police investigation of the incident attributed the cause of the accident to the failure of the passenger car to yield to a responding emergency vehicle. The driver of the passenger car acknowledged that he had seen the responding engine approaching but thought that he could get through the intersection before the engine got there. The report also concluded that Captain Bethune's failure to wear a seatbelt was a major factor in his death. Captain Bethune was the first firefighter fatality for the Texas City Fire Department since most members of the Texas City Fire Department were killed in an explosion 52 years before. Additional information about this incident can be found in NIOSH Fire Fighter Fatality Investigation 99-F-36.

10/7/99

Marvin Huisman, First Assistant Chief
Age 56, Volunteer
Wilmont Fire Department, Minnesota

Chief Huisman and other firefighters were battling a brush fire. Chief Huisman had assisted with the extension of hose lines and was operating a fire pump. He suffered a heart attack and died.

10/7/99
Elvis Benson Maxwell, Firefighter/Operator
Age 49, Volunteer
Grant Parish Fire District #5, Louisiana

Firefighter Maxwell was responding as the driver of a fire department tanker (tender) to a structure fire at night and in the rain. He lost control of the vehicle, left the roadway, and overturned. An EMS unit en route to the fire came upon the scene and discovered Firefighter/Operator Maxwell still inside the vehicle with no vital signs. ALS was administered at the scene and en route to the hospital. Firefighter/Operator Maxwell was pronounced dead in the emergency room. Firefighter/Operator Maxwell was not wearing a seatbelt and was partially ejected in the collision. The cause of death was listed as blunt trauma.

10/16/99
Karen Jane Savage, Firefighter/EMT
Age 44, Volunteer
Hawkins Bar Volunteer Fire Department, California

Firefighter/EMT Savage and other members of her department responded to a wildland fire that had developed into firestorm conditions. The fire eventually consumed 26,000 acres and destroyed 75 homes. Firefighter/EMT Savage and other firefighters stopped at a support vehicle to get supplies. As Firefighter/EMT Savage handed supplies to other firefighters on a pumper, the vehicle began to move. Firefighter/EMT Savage fell or jumped from the pumper and was crushed by the pumper's rear wheels. Additional information about this incident can be found in NIOSH Fire Fighter Fatality Investigation 99-F-42.

10/18/99
Charles C. Young, Firefighter/EMT
Age 77, Volunteer
Ross Township Fire Department, Ohio

Firefighter/EMT Young responded to a very upsetting incident involving the suicide of a teen. He returned from the incident and was speaking with his wife on the phone about the call when he was stricken with a heart attack. His wife was unsure of where he was so it took firefighters 10 to 20 minutes to find and treat him. Despite efforts by members of his department and others, Firefighter/EMT Young died.

10/28/99
Brian K. Burnett, Firefighter
Age 23, Volunteer
Scipio Township Volunteer Fire Department, Indiana

Robert Charles Ulrich, Captain
Age 57, Volunteer
Scipio Township Volunteer Fire Department, Indiana

Firefighter Burnett and Captain Ulrich were responding in a tanker (tender) to the report of a brush fire, Firefighter Burnett who was driving, failed to negotiate a curve in the road; the apparatus left the road and crossed into a cornfield, where it rolled several times. Firefighter Burnett was ejected from the vehicle and the vehicle rolled on top of him. Captain Ulrich was trapped in the tanker, which was on its roof, until he was extricated by other firefighters. Both firefighters were transported to the hospital.

Captain Ulrich died on November 4, 1999. He had been released from the intensive care unit to a regular hospital floor. Captain Ulrich was seemingly well and recovering from his injuries. He was discovered pulseless and non-responsive; emergency care was provided but was not successful. The autopsy concluded that Captain Ulrich died of a cardiac arrhythmia. It is not known if the cardiac problems were related to the collision. Captain Ulrich did have some health factors that contribute to cardiac disease. Captain Ulrich was wearing his seatbelt at the time of the collision.

Firefighter Burnett died on January 22, 2000. He was making a slow recovery. The family had been told that he might be home in a week or so but that he would need further therapy. No cause of death for Firefighter Burnett is available. Firefighter Burnett was not wearing his seatbelt at the time of the collision.

10/29/99
David Merle Pack, Forestry Aide I
Age 63, Wildland Full-Time
Tennessee Department of Agriculture - Forestry Division, Tennessee

Forestry Aide I Pack responded to what was described as a "routine" woodland fire. Other firefighters spoke with him as they responded to the incident. At the conclusion of the incident, Forestry Aide I Pack's pickup truck was found at the edge of a pond near the fire area with the headlights on and the engine running. After a foot search failed to locate Forestry Aide I Pack, a search dog was called in. The dog led searchers to Forestry Aide I Pack's body in the pond. His body was recovered. The reason for Forestry Aide I Pack's presence in the pond is unknown. His cause of death was drowning.

10/29/99
Walter F. Vaughan, Fire Police Officer
Age 80, Volunteer
Warminster Fire Department, Pennsylvania

Fire Police Officer Vaughan was directing traffic around the scene of a reported structure fire. He was struck by a passenger car and sustained multiple injuries. He was transported to the hospital where he was placed on a ventilator. Fire Police Officer Vaughan died on November 13, 1999. At the incident he was wearing a reflective safety vest and helmet and using a wand-type flashlight to direct traffic. The driver of the passenger vehicle was cited for careless driving and failure to obey an authorized person directing traffic.

11/2/99
Michael J. Sims, Sr., Firefighter
Age 38, Volunteer
Highland Hose Company, Tarentum, Pennsylvania

Firefighter Sims was responding to an automatic fire alarm activation as a passenger in a 1965 open cab aerial ladder apparatus. As the truck made a turn, Firefighter Sims fell from the vehicle and sustained severe injuries. His fall was not witnessed, but another firefighter heard the impact of Firefighter Sims striking the pavement. Emergency medical care was provided, and Firefighter Sims was airlifted to a hospital. Firefighter Sims died the following day. The status of his seatbelt was not known. The apparatus was equipped with seatbelts and a safety gate.

11/3/99
Jerry Wayne Ramey, Firefighter Trainee
Age 18, Volunteer
West Fork Fire Department, Arkansas

Firefighter Trainee Ramey responded with members of his department to a fire in the utility easement behind a home. A very small fire was found, which was out except for a few burning embers. Firefighter Ramey attempted to stomp out the embers when he came into contact with a 7,200-volt electrical line which had been hidden from view in tall grass. Firefighter Trainee Ramey fell on top of the line and was removed by other firefighters. Emergency medical aid was provided and Firefighter Trainee Ramey was transported to the hospital where he was pronounced dead. Firefighter Trainee Ramey had joined the fire department after a younger brother had died of an asthma attack at a local high school football game.

11/4/99
William Walter Korte, Firefighter
Age 59, Volunteer
Southampton Fire Department, New York

Firefighter Korte was performing scene safety duties as other members of his department performed a vehicle extrication. Firefighter Korte had just finished closing a road for traffic control. He was standing by the fire police truck talking to another firefighter when he dropped to the ground. Firefighter Korte died of an apparent heart attack. No autopsy was performed.

11/7/99
David Zan Lancaster, Firefighter
Age 24, Volunteer
Elliott Volunteer Fire Department, Mississippi

Firefighter Lancaster was killed as the result of a motor vehicle collision in his personal vehicle while responding to a car fire.

11/14/99
Bert Andrew Bruecher, Firefighter
Age 46, Volunteer
Village of Pleak Volunteer Fire Department, Texas

Firefighter Bruecher was the driver and lone occupant of a tanker (tender) that was responding to a fire involving 200 round bales of hay that were near a home and a propane tank. The tanker entered a curve at high speed, left the road, and rolled over. A shift in the water load may have contributed to the collision. Firefighter Bruecher was partially ejected and was pinned under the truck. Two (2) 14 year-old-boys were arrested and one was charged with second degree arson for setting the fire. As a part of a plea bargain, the boy was placed on probation until he is 18.

11/16/99
Brian Andrew Lee, Firefighter
Age 38, Career
Fire Department Jersey City, New Jersey

Firefighter Lee and his fire company had just returned from an emergency response and were resupplying their engine company. Firefighter Lee began to experience severe stomach pains. An ambulance was called and Firefighter Lee was loaded for transport to the hospital. While en route to the hospital, Firefighter Lee went into cardiac arrest and was not revived. The cause of death was listed as natural, an inflammation of the heart caused by a disease called sarcoidosis.

11/18/99
Henri Fred Broussard, Fire Chief
Age 69, Volunteer
Maurice Volunteer Fire Department, Louisiana

Maurice firefighters responded to a fire that involved the cab of an 8,600-gallon gasoline tanker next to several above ground fuel tanks and a large LP gas tank at a local gas station. Chief Broussard drove the first engine to the scene and was met there by other firefighters. As the firefighters dressed in their protective clothing, Chief Broussard stretched the initial attack line and then returned to the pumper to operate the pump from the top-mounted pump panel. Shortly after arriving back at the pumper, Chief Broussard suffered a heart attack. Immediate medical aid was provided by other firefighters and the crew of an on scene ambulance. Chief Broussard was transported to a local hospital where he was pronounced dead upon arrival. No autopsy was performed. 3 other firefighters were injured at the incident.

11/18/99
James Melvin Dunham, Safety Officer/Firefighter
Age 36, Volunteer
Saint Jo Fire Department, Texas

Firefighter Dunham drove a fire department rescue truck to the scene of a mutual-aid vehicle collision that required extrication. As he was setting up the power unit for a hydraulic rescue tool, Firefighter Dunham was seen to stumble and hit his head on the ground. Other firefighters rendered immediate aid, including the use of an AED. Advanced life support was provided on the scene and Firefighter Dunham was flown to a hospital by helicopter. Despite all efforts, Firefighter Dunham did not survive. The cause of death was listed as "occlusive coronary artery atherosclerosis."

11/20/99
Jackie Mac Garnett, Firefighter
Age 54, Volunteer
Quapaw Volunteer Fire Department, Oklahoma

Firefighter Garnett and his department were dispatched to a wildland fire. Firefighter Garnett rose to respond and suffered a heart attack. He was transported to a local hospital, stabilized, and then flown to a regional hospital. He was taken to surgery and died a short time later. No autopsy was performed.

11/20/99
Alton L. "Al" Lewis, First Assistant Fire Chief
Age 55, Volunteer
Montour Falls Fire Department, New York

First Assistant Fire Chief Lewis had just returned home after responding to a vehicle collision with a building. He suffered a heart attack in his driveway and died.

11/20/99
Wayne C. Yost, Assistant Fire Chief
Age 48, Volunteer
Cochranville Fire Company, Pennsylvania

Assistant Chief Yost had responded to a shed and wildland fire. He complained of not feeling well at the scene and went home. Shortly after his arrival at home, an ambulance was called. Assistant Chief Yost had suffered a heart attack. He was brought back at his home by members of a local response team. Assistant Chief Yost died on November 27, 1999.

12/2/99
Brad A. Michener, Firefighter
Age 24, Volunteer
Scipio-Republic Fire Department, Ohio

Firefighter Michener had driven a heavy rescue truck to the scene of a brush fire and then back to the station at the conclusion of the incident. Firefighter Michener left the station in his personal vehicle and returned home. His home was about a block from the fire station. As he exited his vehicle at home, he collapsed in his backyard of an apparent heart attack. Firefighter Michener died the next day, December 3, 1999. Firefighter Michener had a pre-existing heart condition but had been released to full duty by his personal physician.

12/3/99

Paul Arthur Brotherton, Firefighter
Age 41, Career
Worcester Fire Department

Timothy Paul Jackson, Firefighter
Age 51, Career
Worcester Fire Department

Joseph T. McGuirk, Firefighter
Age 38, Career
Worcester Fire Department

Jeremiah Michael Lucey, Firefighter
Age 38, Career
Worcester Fire Department

James Francis Lyons, Firefighter
Age 34, Career
Worcester Fire Department

Thomas Edward Spencer, Lieutenant
Age 42, Career
Worcester Fire Department

Members of the Worcester Fire Department responded to a fire in the Worcester Cold Storage Warehouse. The building was a windowless 6-story structure. Upon arrival at the scene, firefighters found a large warehouse with light smoke conditions and a fire on the second floor. Search and rescue and fire attack operations were initiated. Within seconds, conditions in the fire building changed and thick black smoke reduced visibility to zero. All fire department personnel were ordered down from upper floors and a head count was taken. The head count revealed that 2 firefighters were not accounted for. A "Mayday" radio transmission was received from Firefighter Brotherton indicating that he and Firefighter Lucey, both of Rescue One, were lost and running out of air. A search for the trapped firefighters was initiated with 18 firefighters searching for the 2 that were lost. Lieutenant Spencer, Firefighter Jackson, Firefighter McGuirk, and Firefighter Lyons entered the fifth floor to conduct a search. Contact with the team was lost and all 6 firefighters perished. Although it has not yet been released, NIOSH is conducting a review of this incident. The cause of the fire is believed to be accidental, the result of a candle knocking over during a domestic dispute by some transients living in the building. The transients were charged with manslaughter for failing to report the fire.

12/7/99
Roy Kenneth Crago, Firefighter
Age 65, Volunteer
Fallston Volunteer Fire and Ambulance Company, Inc., Maryland

Firefighter Crago and members of his fire department were dispatched to a report of a transformer explosion. Firefighter Crago arrived at the fire station in his personal vehicle and responded as the driver of an engine company apparatus. Two (2) other firefighters were on board. At some point during the response, Firefighter Crago suffered a stroke. The engine left the road, skidded down an embankment, and crashed into a concrete culvert. Other firefighters removed Firefighter Crago from the apparatus and began CPR. ALS was provided by other firefighters and Firefighter Crago was transported to the hospital. Despite all efforts, Firefighter Crago died on December 10, 1999. The cause of death was listed as a subarachnoid hemorrhage caused by a brain aneurysm. Firefighter Crago was not wearing his seatbelt, but the injuries he received as a result of the crash were not life threatening.

12/10/99
Richard L. Van Wert, Fire Chief
Age 58, Volunteer
Schaghticoke Fire Department, New York

Chief Van Wert was supervising the disposal of fireworks residue in a controlled burn at a local fairgrounds. Chief Van Wert noticed a spark heading toward a van containing at least 100 pounds of additional residue that was to be disposed of. He yelled for the fireworks worker to run but was unable to escape the explosion himself. The van exploded and burned. Chief Van Wert was killed instantly. Chief Van Wert's actions were credited with saving the fireworks company employee.

12/13/99
Gregory Eugene Rodgers, Firefighter/EMT
Age 50, Volunteer
Dresden Volunteer Fire Department, Ohio

Firefighter Rodgers responded as the passenger of a water tanker (tender) that responded to a mutual-aid barn fire. The driver of the tanker was his son, a Firefighter/EMT. Firefighter Rodgers assisted with the setup of a portable tank and suction equipment and helped the tanker dump its load of water at the fire scene. Once empty, the tanker responded to a hydrant about 2 miles from the fire scene and connected to it. Firefighter Rodgers was found on the ground and unresponsive by the driver of the tanker. ALS was called from the fire scene, and the driver began CPR. Firefighter Rodgers was transported to the hospital but succumbed to a heart attack.

12/15/99

Paul Franklin Ezernack, Jr., Firefighter

Age 28, Volunteer

North Sabine Fire Protection District, Louisiana

Firefighter Ezernack was responding in a 1,500-gallon tanker (tender) to a report of a brush fire. An embankment gave way under the right wheels of the tanker. Firefighter Ezernack attempted to regain control, but the tanker left the roadway and rolled over. Firefighter Ezernack was ejected from the vehicle and was reported to have been thrown 170 yards. Firefighter Ezernack was pronounced dead at the scene.

12/18/99

Bradley Curtis McNeer, Firefighter

Age 22, Volunteer

Chesterfield County Fire Department, Virginia

Firefighter McNeer was riding in the officer's seat of a heavy rescue responding to a gas leak in a residence. He and the driver were the only occupants of the vehicle. Neither firefighter was familiar with the route to the incident, and Firefighter McNeer was having difficulty finding the address in the apparatus map book. The driver decided to reduce his response mode to non-emergency and pull over to look at the map book himself. As he was preparing to pull over or stop, the right rear wheels of the apparatus left the road and went into a ditch. The driver was able to steer the unit out of the ditch, but oversteered to the left and struck a car. The driver then saw another oncoming vehicle and overcorrected to the right ending up back in the ditch. The heavy rescue struck a large tree. Firefighter McNeer was wearing a seatbelt but the severity of the crash was too great. Firefighter McNeer received a fatal head injury. A fire department investigation concluded that "The accident was caused by the driver taking his eyes off the road as he reached for the light switch. Attempting to drive the vehicle out of the ditch, and the speed of the vehicle, contributed to the severity of this accident."

12/20/99

John H. Tvedten, Battalion Chief
Age 47, Career
Kansas City Fire Department, Missouri

Chief Tvedten was a sector officer working inside a warehouse that was involved in fire. Visibility in the warehouse was good, and firefighters were putting water on the fire. About 45 minutes into the incident, interior conditions changed rapidly as thick black smoke enveloped the building. Command ordered the building to be evacuated and Chief Tvedten ordered firefighters to leave. The emergency evacuation signal was given over radios and by fire apparatus air horns at the scene. During the evacuation, Chief Tvedten became disoriented and lost. Chief Tvedten was in radio communication with Command. 6 search teams swept the building but were not able to locate Chief Tvedten until it was too late.

12/20/99

Theodore A. Ferrante, Jr., Firefighter
Age 43, Career
Revere Fire Department, Massachusetts

Firefighter Ferrante and members of his ladder crew responded to an alarm of fire in a highrise building. The cause of the alarm was found to be a prank pull station activation. At the scene of the incident, Firefighter Ferrante complained of chest pains but told his company officer that he just needed to lie down and he would feel better. Approximately 2 hours later, Firefighter Ferrante began to experience severe pain. An ambulance was called as other firefighters rendered aid. Firefighter Ferrante was transported to a local hospital where he died in the early morning hours of December 21, 1999. The cause of death was a heart attack.

12/22/99

Jason L. Bitting, Firefighter
Age 29, Career
Keokuk Fire Department, Iowa

David M. McNally, Assistant Chief
Age 48, Career
Keokuk Fire Department, Iowa

Nathan R. Tuck, Firefighter
Age 39, Career
Keokuk Fire Department, Iowa

The Keokuk Fire Department was dispatched to a fire in a residential structure. The structure was a house built in 1910 that had been divided into three apartments. The department responded with an engine, a quint, and a Chief's vehicle with a total of 3 Firefighters, a Lieutenant, an Assistant Chief, and the Fire Chief. The response of the Chief and one Firefighter was delayed because they were returning from a previous incident. Upon arrival, Assistant Chief McNally, Firefighter Bitting, and Firefighter Tuck entered the building in full turnouts and SCBA for search and rescue. A mother and child were trapped on the roof above the porch and 3 other children were trapped inside. Firefighters rescued 1 infant child who was transported to the hospital by a police officer. Firefighters rescued a second infant child who was transported to the hospital by a police Captain and the Fire Chief. The Fire Chief was away from the scene for approximately 3 minutes. Firefighters were searching for the third child when a flashover occurred and trapped all 3. An aggressive fire attack was mounted by firefighters who were arriving as part of a callback of off-duty members but the effort was not able to save the lives of the 3 firefighters.

All 3 firefighters were wearing PASS devices that were integrated with their SCBA's. The PASS devices failed to sound an alarm when the firefighters became incapacitated. The Fire Chief does not believe that the failure of the PASS devices contributed to the deaths. The SCBA's and PASS devices are undergoing testing to determine why they did not operate.

Assistant Chief McNally was found on the second floor at the top of the stairs with the third child. The cause of death was listed as smoke inhalation and exposure to extreme heat. Chief McNally's carboxyhemoglobin level was 15 percent. Firefighter Bitting was found in the front bedroom on the second floor of the apartment. The cause of death was listed as exposure to intense heat. Firefighter Bitting's carboxyhemoglobin level was 1 percent. Firefighter Tuck was found on the first floor of the apartment in the living room area. The cause of death was listed as smoke inhalation and exposure to heat. Firefighter Tuck's carboxyhemoglobin level was 25 percent.

In addition to the 3 firefighters killed in this incident, the 2 infant children and a 7-year-old child perished. The fire was caused by activation of a stove burner by a child. Two (2) high chair trays that were stored on top of the stove were the initial objects involved in the fire. Smoke detectors in the home did not operate.

12/29/99

Ronald Eugene Kaltreider, Safety Officer

Age 39, Volunteer

Pleasant Hill Volunteer Fire Company, Pennsylvania

Safety Officer Kaltreider had been at the fire station for most of the day performing year-end computer work and assuring that his department was prepared for Y2K. As he discussed the upcoming purchase of some communications equipment with another firefighter, he suffered a heart attack. Firefighters immediately began attempts to revive Firefighter Kaltreider, but they were unsuccessful. Firefighter Kaltreider had a history of heart disease dating back to 1990.

12/31/99

Robert Dale Pollard, Firefighter

Age 64, Volunteer

Southern Stone County Fire Protection District, Missouri

Firefighter Pollard was driving a rescue vehicle to a wildland fire. While en route, he collapsed and was treated by first responders and then airlifted to a hospital. Firefighter Pollard died the next day on January 1, 2000, of a cerebral bleed (stroke/CVA).

Additional information about many of the firefighter fatalities presented in this appendix is available from the sources below. Where known, the report number for each incident is listed in the appendix along with the incident description. Many reports are available through the mail and the Internet.

Kentucky Division of Forestry
502-564-4496
http://www.nr.state.ky.us/nrepc/dnr/forestry/dnrdof.html

National Fire Protection Association
1 Batterymarch Park
P.O. Box 9101
Quincy, MA 02269
(617)-770-3000
http://www.nfpa.org

National Institute for Occupational Safety and Health (NIOSH)
Fire Fighter Fatality Investigation and Prevention Program
1095 Willowdale Road
Mail Stop P-180
Morgantown, WV 26505-2888
http://www.cdc.gov/niosh/firehome.html
(800)-35-NIOSH

Manteca Fire Department, California
http://www.ci.manteca.ca.us/fire/bluegum.html

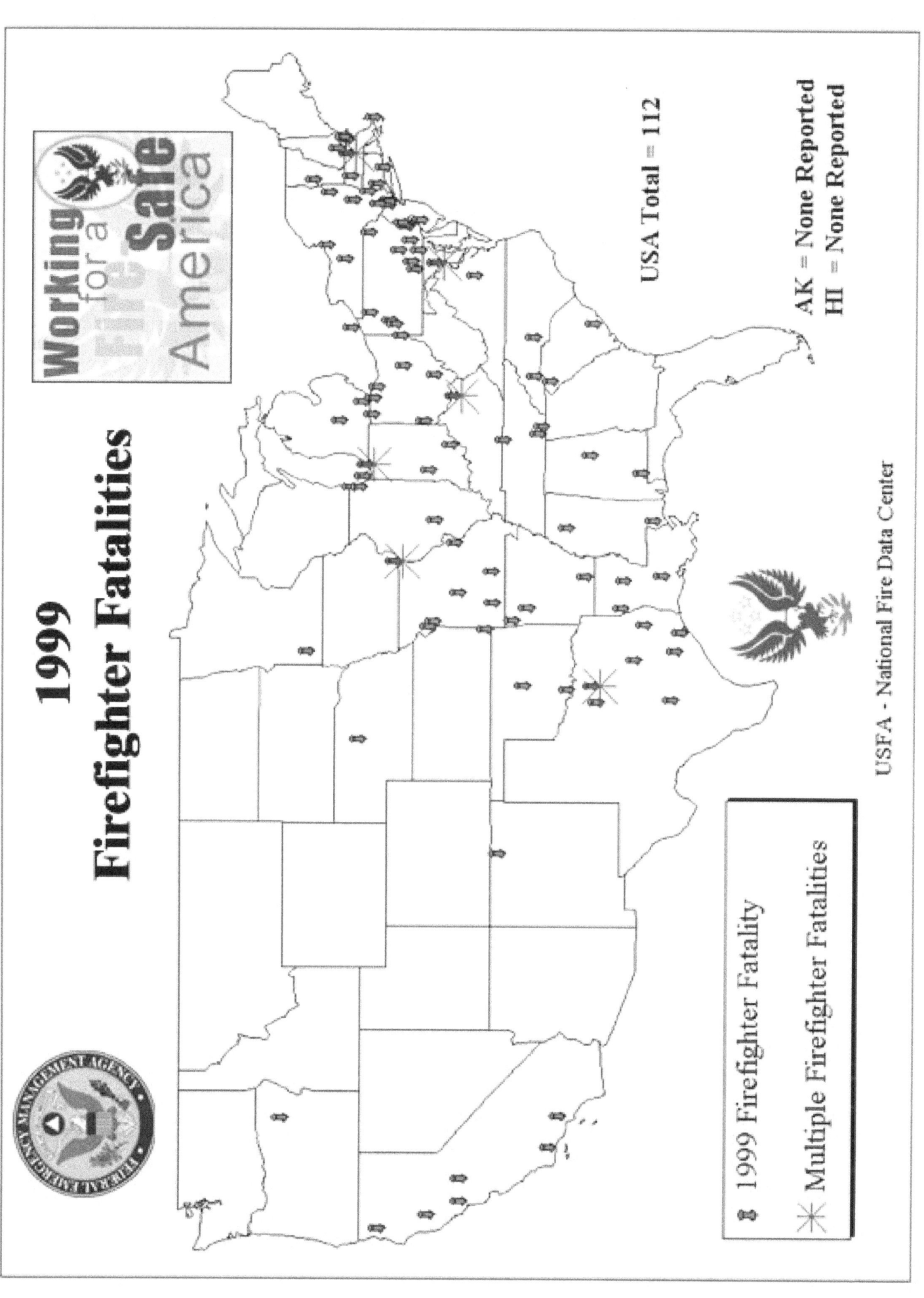

1999
Firefighter Fatalities

USA Total = 112

AK = None Reported
HI = None Reported

USFA - National Fire Data Center

Working for a Fire Safe America

🔥 1999 Firefighter Fatality

✳ Multiple Firefighter Fatalities